HD
9566
.S497   Shaffer, Ed.
1983
         The United States
         and the control of
         world oil

DATE			

THE UNITED STATES AND THE CONTROL OF WORLD OIL

ED SHAFFER

ST. MARTIN'S PRESS   NEW YORK

Library of Congress Cataloging in Publication Data

Shaffer, Ed.
   The United States and the control of world oil.

   Bibliography: p.
   Includes index.
   1. Petroleum industry and trade – Government policy –
United States. 2. Petroleum industry and trade –
Political aspects – United States. I. Title.
HD9566.S497   1983   338.2'7282'0973   82-42720
ISBN 0-312-83314-8

CONTENTS

To Florence, Martha and Paul

TABLES

ACKNOWLEDGEMENTS

In writing this book I have received suggestions, aid and comments from many people, too many to mention in a brief acknowledgement. Nevertheless, I would like to express my thanks to several individuals who were particularly helpful. Among them are Dr Larry Pratt, Department of Political Science, University of Alberta and Mr David Leadbeater, presently a graduate student at Massey College, University of Toronto, who directed me to some important sources of information. I would also like to thank Dr Gilbert Reschenthaler, Faculty of Business Administration and Commerce, University of Alberta, who made many valuable comments on the opening chapters. I want to render special thanks to Dr Clement Leibovitz, Computing Services, University of Alberta, for the many hours he spent helping to prepare the manuscript for the computer. Finally, I would like to thank Energy, Mines and Resources Canada for the financial assistance it gave me in preparing this book. Needless to say, the opinions expressed are mine entirely and do not necessarily reflect the opinions of those who helped me.

1 INTRODUCTION: THE AGE OF OIL

A spectre is haunting the world — the spectre of an oil famine. The major powers of the world have entered into a holy alliance to exorcise this spectre; but to no avail. The spectre will continue to haunt the world for many years.

This adaptation of Marx and Engels[1] aptly describes the fears of the major industrialised market societies in the latter part of the twentieth century. The events of the 1970s have shaken the very foundations of these societies. In those years the world was emerging from an era in which oil was both cheap and readily available into one in which it has become dear and relatively scarce.

This new era is forcing these societies to make many painful adjustments. These adjustments are both of a physical and a social nature. The physical involves finding more efficient ways to use oil and developing substitutes for it. The social implies major shifts in the distribution of world income and concomitant changes in world power relationships.

Of the two adjustments, the social is undoubtedly the more complex. In comparison, the physical is relatively simple. Science and technology will find ways to use oil more efficiently as well as to discover substitutes for it. They will not, however, be able to deal with the social disturbances that will accompany the end of the oil era.

The task of predicting, analysing and offering possible solutions to these upheavals should fall to the social scientists. Unfortunately, today's social scientists, and particularly economists, are not equipped to handle this task. Economists have abandoned the tradition of political economy in favour of an allegedly more abstract, value-free 'positive' economics. However, the concept of a value-free science is an absurdity. All scientists are products of the value systems of the societies in which they live. The choice of research paths in both the physical and social sciences always reflects these systems. More important, the pursuit of a value-free science has caused economists to concentrate on testing narrowly-defined hypotheses. The advent of the computer has further led them to test only those hypotheses that are readily quantifiable.[2] In doing so they have avoided raising basic questions about the social structures of modern societies. They have accepted these structures as given.

Though economists like Adam Smith, Karl Marx and Joseph

Schumpeter recognised relationships between technical and social change, modern economists have by and large ignored these relationships. It is not surprising therefore that their analyses of the world's economic behaviour in the 1970s were, as in the 1930s, way off the mark. As Wassily Leontieff remarked in his presidential address to the American Economic Association, 'In no other field of empirical inquiry has so massive and sophisticated a statistical machinery been used with such indifferent results.'[3]

It is for this reason that this book proposes to examine the problems posed by the rise and fall of oil in the tradition of political economy. It particularly analyses the relations between technical and social change as they apply to oil. Only through an understanding of these relationships can one comprehend the implications of the rise and fall of oil.

In Schumpeterian terms the ushering-in of the oil era was an example of 'the perennial gale of creative destruction'.[4] The creative winds of the new product, oil, destroyed the old one, coal. But this 'perennial gale' does something more than replace an old product with a new one. It also displaces the established entrepreneurs, i.e. those making the old product, with a new group of entrepreneurs, i.e. those introducing the new product.[5] The rise of this new group brings a change in the power relationships of the economy. Furthermore, on an international scale, it may bring a change in the power relationships among nations. These changes in international power relationships did not concern Schumpeter. They are however central to the theme of this book.

Its theme is that the rise of oil in the United States gave birth to a new group of entrepreneurs, the international oil companies, who began to play an increasingly important role in the domestic economy. Furthermore, the very nature of oil, a commodity whose supply is finite and whose geographical distribution is highly unequal, has led these companies to expand in other countries in their search for oil. Their expansion could have not been successful without the active help of the US government, which pursued a deliberate policy of acquiring domination over the world's oil. The rise of oil hence coincided with the rise of the American Empire and with the attempts by the US to establish hegemony over the rest of the world.

New forces are now at work which are undermining America's position. Among them is the inability of the US to control oil in many formerly 'secure' areas. Not only is this inability evident in the OPEC (Organisation of Petroleum Exporting Countries) nations but it is also evident in Mexico and in Canada, where a struggle is going on to free their resources from US control.

In addition, the end of the oil era may further weaken the US position. There is no guarantee that America will be able to control the new energy sources and the new technology necessary to make them operable. To retain its dominant position the US will be forced to wage a relentless struggle to control the energy sources and the new technology. This struggle will affect all of the world's industrialised countries, and especially the market economies, as they engage in a race with the US to be the first to develop the new technologies. One aspect of this race is the introduction in all the major market economies of austerity programmes, which are designed to shift a greater share of national income from consumption into investment in energy projects.

This book intends to analyse this struggle, this new form of economic warfare. It will do so by tracing (1) the rise of oil in the United States, (2) the movement of the US to a position of world dominance, and (3) its decline from this position.

Notes

1. Karl Marx and Friedrich Engels, *The Communist Manifesto* (International, New York, 1980), p.8.

2. Economists often try to take account of qualitative variables through the use of dummy variables. While useful in some types of research, this technique usually can give only a superficial understanding of the relationships between quantitative and qualitative variables.

3. W. Leontieff, 'Theoretical Assumptions and Unobserved Facts', *American Economic Review*, vol. 61, no. 1 (1971), p.3. For other criticisms of modern economics see J. Robinson, 'The Second Crisis of Economic Theory', *American Economic Review*, vol. 62, no. 2 (1972), pp.1-10 and J.K. Galbraith, 'Power and the Useful Economist', *American Economic Review*, vol. 63, no. 1 (1973), pp.1-11.

4. J. Schumpeter, *Capitalism, Socialism, and Democracy*, 2nd edn. (Harper and Brothers, New York and London, 1947), p.84.

5. Schumpeter's thesis that new entrepreneurs are associated with new products is only partially true. While it does portray the pattern in oil, it is inapplicable in other industries. Innovations have also been introduced by established firms, e.g. nylon by DuPont and the transistor by Bell Labs.

2 THE EARLY EXPANSION OF AMERICAN CAPITALISM

From its birth in 1776 the United States has pursued expansionist policies. This is not surprising since the American Revolution occurred during the Mercantilist Era, an era in which all the major powers were also pursuing such policies. While the Revolution has often been interpreted as a revolt against mercantilism in general, it was in actuality a revolt against the application of specific British mercantilist policies to the North American colonies.

One of these was the attempt by the British to reserve the western frontier areas to the fur trade. This attempt clashed with the desire of settlers and land speculators to bring these lands under their control. It was thus inevitable that, with the ousting of the British, the fledgling American state would emulate its European rivals in a policy of expansion.

The aggressiveness of the young Republic can be shown through the relatively rapid growth of the territory under its control. In 1783, the year in which Britain signed the peace treaty with the US, its area consisted of approximately 500,000 square miles. By 1853, 70 years later, the country extended from 'sea to shining sea', encompassing an area of over three million square miles.[1] After acquiring the central part of the North American continent, the nation's interests turned outward.

Continental Consolidation

The conquest of the continent was not a peaceful one. It required three major wars (the War of 1812, the Mexican War and the Civil War) and a host of minor ones against the Indians. The latter, though not on the scale of the major conflicts, were nevertheless bloody and costly.[2] All these wars had the effect of spreading capitalism across the continent, thereby creating the basis for the expansion of the US industrial machine.

The Civil War introduced capitalism into the South. The North's victory enabled its industrialists to expand their operations in that region. As the freed slaves entered the market economy, they provided both a market for manufactured goods and a relatively cheap source of labour. The growth of a market in the South encouraged the

4

building of railroads, further integrating it into the national economy. The *ante-bellum* South, it should be noted, had stronger economic links with England than with the North. Because of these links, Britain supported the Confederacy during the Civil War. The weakening of these links gave an important impetus to the expansion of Northern industry.

The Mexican and Indian wars enabled the country to conquer the West. As a result of the Mexican War the nation added more than 800,000 square miles to its territory.[3] This territory, which includes, among others, the States of Texas, California, New Mexico, Arizona, Nevada and Colorado, contains some of the richest mineral deposits in the world. The acquisition of two of these states, Texas and California, was instrumental in enabling the United States to establish itself as a major oil power.

The decimation of the Indians in the Indian wars[4] opened this vast territory to settlement and exploitation. With settlement and exploitation came the transcontinental railroads, which linked these areas to the industrial East. The railroads however did more than merely provide economic links between East and West. They provided the basis for a major restructuring of the US economy.

By integrating the economies of the South and West with the rapidly industrialising North-East, they created one national market. A unified common market displaced a series of disparate regional ones. The new national market gave birth to national firms, which either bought out or destroyed the small-scale regional enterprises that characterised the pre-Civil War economy. The large company became the symbol of American capitalism.

In the period between the end of the Civil War and World War I, America witnessed the greatest consolidation movement in its history. These consolidations took the forms of trusts, holding companies and mergers.[5] This period gave birth to such giant firms as United States Steel, American Tobacco, International Harvester, Anaconda, American Sugar Refining, Armour, Standard Oil of New Jersey, the large railroads and many other giant corporations.[6] All these companies acquired many plants, mines and establishments. Through these acquisitions they achieved dominant positions both in their respective markets and in the economy as a whole. Table 2.1 lists the number of plants controlled by some of the major firms at the turn of the century.

This form of growth, i.e. expansion through the acquisition of control of production facilities in diverse geographic regions, represented 'the wave of the future'. It became inevitable that this growth would

Table 2.1: US Plants Controlled by Major Corporations, 1901

Corporation	Plants (number)
Amalgamated Copper Co.	11
American Smelting and Refining Co.	121
American Sugar Refining Co.	55
Consolidated Tobacco Co.	150
Standard Oil Company of New Jersey	400
United States Steel Corporation	785

Source: J. Moody, *The Truth about the Trusts* (Moody Publishing, Chicago, 1904), p.453.

breach the confines of the domestic economy and spill over into foreign ones. With the consolidation of the central portion of the continent completed, the American state turned its eyes towards other areas.

External Expansion

Shortly after the end of the Civil War in 1865, the United States began to consider seriously expansion beyond the mid-continent. It made its first important moves in 1867, when it annexed the Midway Islands in the Pacific[7] and purchased Alaska from Russia.[8] The next major step was the Spanish-American War of 1898. As a result of that war the US acquired the following territories: Puerto Rico, the Philippines, Guam, Hawaii and Samoa. The first three were seized from Spain. Hawaii, which was nominally independent at the outset of the war, was formally incorporated into the United States, while Samoa, which had a protectorate status at the outbreak of hostilities, was ceded to the United States by Germany and Great Britain, who also had claims to that island chain.[9] In the years following the Spanish-American War the US acquired control of the Panama Canal Zone, as a result of a treaty imposed on Panama in 1903, and the Virgin Islands, which it purchased from Denmark in 1917. It also obtained in 1914 a 99 year lease on the Corn Islands from Nicaragua.[10] These acquisitions are shown in Table 2.2.

Despite these acquisitions the United States possessed a relatively small empire. Its overseas territories contained only 700,000 square miles, of which almost 600,000 were in the vast emptiness of Alaska.[11] In comparison to the empires of England and France, its overseas territory was relatively insignificant. Nevertheless the United States

Table 2.2: Overseas Territorial Expansion of the United States, 1867-1917

Territory	Year acquired	Area (000 sq. miles)	Method of acquisition
Alaska	1867	586	Purchased from Russia.
Midway Islands	1867	a	Formally annexed. First reported and claimed for US in 1859 by Navy Capt. N.C. Brooks.
Hawaii	1898	6	Independent kingdom seized by US.
The Philippines	1898	116	Taken from Spain as a result of Spanish-American War.
Puerto Rico	1899	3	Taken from Spain as a result of Spanish-American War.
Guam	1899	a	Taken from Spain as a result of Spanish-American War.
Samoa	1900	a	From treaty with England and Germany.
Panama Canal Zone	1904	1	From treaty imposed on Panama.
Corn Islands	1914	a	Leased from Nicaragua for 99 years in treaty imposed on that country.
Virgin Islands	1917	a	Purchased from Denmark.
Total		712	

Note: a. Less than 500 square miles.
Sources: *Encyclopaedia Britannica* (Chicago, 1962), vol. 1, p.501 and vol. 15, p.460; *Historical Statistics of the United States, Colonial Times to 1957: A Statistical Abstract Supplement* (US Government Printing Office, Washington, 1960), p.236; S. Nearing and J. Freeman, *Dollar Diplomacy*, reprint (Monthly Review Press, New York and London, 1966), p.283.

became a major world power during this period because it was able to exercise hegemony over most of the countries in the Western Hemisphere without formally incorporating them in the Union. This policy of rule without incorporation was best exemplified by the treatment of Cuba.

Acquisition Without Annexation

Instead of annexing Cuba, the United States gave that country formal independence. At the same time it forced the Cubans to incorporate the Platt Amendment into their Constitution. This amendment, which was introduced into the US Congress by Senator Platt of Connecticut, granted the United States 'the right to intervene for the preservation of Cuban independence, the maintenance of a government adequate for the protection of life, property, and individual liberty' and stipulated that 'the Government of Cuba will sell or lease to the United States lands necessary for coaling or naval stations at certain specified points, to be agreed upon with the President of the United States'.[12] In 1903 the United States used this latter stipulation to obtain a naval base at Guatanamo,[13] which it has occupied ever since.

There were a number of reasons why the US decided not to annex Cuba. First, it did not want the responsibility for caring for the Cubans. Second, it felt it could obtain the results it wanted, i.e. economic domination, without going through the costly process of formal colonisation. In fact during the Spanish-American War a debate raged over the necessity of incorporating any new territories in the Union. Not only was there opposition to incorporating Cuba but to incorporating Hawaii and the Philippines as well. The latter two were incorporated only because they were regarded as essential stepping-stones to the potentially lucrative trade with China. Their incorporation was however to be perceived as the exception to the general rule of non-annexation.

This philosophy towards formal annexation was perhaps best expressed by Carl Schurz, one of the country's most influential political leaders and a man who probably would have been President had he not been born in Germany.[14] In a famous article, 'Manifest Destiny', written for *Harper's* in 1893, he argued against colonialism of any kind, stating that the US could gain its economic objectives 'without taking those countries into our national household . . . without assuming any responsibilities for them'.[15] The Secretary of State at that time, Walter Gresham, commented that this was 'the best article of the kind I have ever seen'.[16] Several years later, at an 1898 meeting of the National Civic Federation, Schurz spelled out more clearly what he meant. In commenting on that speech W.A. Williams notes:

> Schurz . . . had always accepted the necessity of market expansion while opposing traditional colonialism. Exports, he had long argued,

'will save us from the suffocation so much dreaded . . . I fully agree . . . we cannot have too many'. But the United States could obtain the required bases and ports 'without burdening itself with any political responsibilities in the regions concerned'. Hence the proper policy was to occupy the outposts 'until they are thoroughly pacified'; and then, having obtained the required economic and military footholds, withdraw. That would extend freedom by 'exerting civilizing influences upon the population of the conquered territories', and 'gain commercial opportunities of so great a value that they will more than compensate for the cost of the war'.[17]

Dollar Diplomacy

This policy of using American military might more for market expansion and commercial gain than for empire building in the traditional sense was reinforced by the adoption of the 'Open Door' policy towards China. Under that policy, the US demanded that the Great Powers allow it equal access to the markets of China. The US also engaged in numerous landings in and temporary occupations of Latin American countries in order to ensure the protection of its economic interests.[18] It pursued a policy of 'Dollar Diplomacy', whose essence was summarised in a 1912 speech by President Taft, who stated:

> The diplomacy of the present administration has sought to respond to modern ideas of commercial intercourse. This policy has been characterized as substituting dollars for bullets . . . It is an effort frankly directed to the increase of American trade upon the axiomatic principle that the Government of the United States shall extend all proper support to every legitimate and beneficial American enterprise abroad . . . If this Government is really to preserve to the American people that free opportunity in foreign markets which will soon be indispensable to our prosperity, even greater effort must be made. Otherwise the American merchant, manufacturer, and exporter will find many a field in which American trade should logically predominate preempted through the more energetic efforts of other governments and other commercial nations . . . We need legislation enabling the members of the foreign service to be systematically brought into direct contact with the industrial, manufacturing, and exporting interests of this country in order that American business men may enter the foreign field with a clear perception of the exact conditions to be

dealt with and the officers themselves may prosecute their work
with a clear idea of what American industrial and manufacturing
interests require . . .[19]

While Taft's address clearly reflected the ideas of Carl Schurz, its
emphasis was somewhat different. Schurz was mainly concerned
with using US power as a means of opening up foreign markets for
US goods. Taft extended this concept to the protection of
'American enterprises abroad.' This change in emphasis was
important because, by the time Taft had assumed the Presidency, the
export of capital had become increasingly important for the US
economy.

During the nineteenth century America was a net importer of
capital. It had to export goods to pay for these capital imports. It
was for this reason that American foreign policy was initially directed
at preserving and expanding markets for American exports. But by
the time continental consolidation had been achieved, and especially
after the large national corporations had emerged, the US began to
export capital as well as goods. The structural changes within the
domestic economy led to changes in the composition of exports,
with the export of capital becoming increasingly important. This
shift to the export of capital, a shift that has occurred in all the
major industrial market economies, has been described by both
Hobson and Lenin as one of the essential features of modern
imperialism.[20]

The shift to significant capital exports probably occurred in the
US during the 1890s. In that period net capital flows changed from
a positive to a negative item in the balance-of-payments account.
In 1895, for instance, the US recorded a net capital inflow of
$137 million. This inflow was reduced to $40 million in 1896 and
then changed to a net outflow of $23 million. This net outflow
jumped sharply to $279 million in 1897 and remained well above
$200 million for the rest of the decade.[21]

Not all this net outflow represented US investment overseas. Prior
to 1900 the capital account was used as a balancing item, i.e. errors
and omissions were placed in it. In addition some of the outflow was
caused by the repatriation by foreigners of investments in the US.[22]
Nevertheless, US overseas investment must have also grown rapidly
in these years because, as Table 2.3 indicates, the US accounts
recorded an investment income of $38 million for 1900. Presumably
most of that income came from investment in previous years.

Table 2.3: US Exports, Direct Investments and Investment Income, 1900-1913

| Year | ($ millions) | | | (Index, 1900 = 100) | | |
	Exports[a]	Invest-[b]ments	Income[c]	Exports	Invest-ments	Income
1913	2,816	138	137	167	246	361
1912	2,738	139	123	162	248	324
1911	2,405	95	114	143	170	300
1910	2,160	124	108	128	221	284
1909	2,013	88	100	119	157	263
1908	2,022	48	89	120	86	234
1907	2,192	89	87	130	159	229
1906	2,052	92	86	122	164	226
1905	1,859	46	76	110	82	200
1904	1,657	80	70	98	143	184
1903	1,663	81	67	99	145	176
1902	1,550	65	57	92	116	150
1901	1,651	89	47	98	159	124
1900	1,686	56	38	100	100	100

Notes: a. Exports of goods and services. b. Private investments only. c. Includes income from both direct and *rentier* investments.
Sources: *Historical Statistics of the United States, Colonial Times to 1957: A Statistical Abstract Supplement* (US Government Printing Office, Washington, 1960), pp.562, 564.

Furthermore, the official US accounts recorded, for the first time, data on US capital outflows for 1900,[23] indicating that they reached such a high level by that year that they warranted entering as a separate item.

Direct Investment

Most of this outflow was direct, as opposed to *rentier*, investment. Direct investment gives the investor control over resource use while *rentier* normally does not. Direct investors are mainly large corporations who seek to control productive facilities in other countries. *Rentier* investors are normally wealthy individuals or large financial institutions who seek to maximise income from a portfolio. They are usually more interested in shifting assets within their portfolio in order to obtain a maximum yield than in acquiring control over a particular company since the acquisition of such control may hinder their flexibility in shifting investments from one company to another.

As Table 2.3 shows, US direct investments grew at a much faster rate than exports between 1900 and 1913, the last full year of peace before the outbreak of World War I. Income from investments, including *rentier* as well as direct, rose at a much faster pace than either exports or investments. Presumably most of this increased income came from the direct, as opposed to the *rentier*, investments.

The reason for this presumption is that direct investments can occur only in a world of imperfect competition, i.e. in a world dominated by large firms which usually sell differentiated products. As Kindleberger notes:

> . . . in a world of perfect competition of goods and factors, direct investment cannot exist. In these conditions, domestic firms would have an advantage over foreign firms in the proximity of their operations to their decision making centers, so that no firm could survive in foreign operation. For direct investment to thrive there must be some imperfection in markets for goods or factors, including among the latter technology.[24]

Direct investment enables large firms to exploit these market imperfections by earning a higher average rate-of-return.

As was pointed out earlier in this chapter, the domestic expansion of the large US firms was characterised by the acquisition of productive facilities in diverse geographical areas. This direct investment at the domestic level became the international model by the turn of the century. In 1897, for instance, the book value of US private direct foreign investment was estimated at $600 million. By 1908 this value had risen by almost three-fold to $1.6 billion (thousand million). By 30 June 1914, on the eve of World War I, the estimated book value had increased to $2.7 billion,[25] more than a four-fold increase over the 1897 level. As can be seen in Table 2.4 at the outbreak of World War I, the bulk of this investment, over two-thirds, was in four industries: mining, manufacturing, agriculture and petroleum. Table 2.5 indicates that almost three-fourths of this investment was concentrated in the Western Hemisphere, an area over which the US exercised hegemony. Europe, the only other area receiving substantial investments, was, at that time, outside the area of US control.

Most of the investment in Europe was in manufacturing, chiefly in those industries in which US firms had either a technological advantage or a differentiated product. Among the firms with

Table 2.4: Distribution of US Direct Private Foreign Investments by Industry, 1914

Industry	Investment ($ millions)	(%)
Mining and smelting	720	27
Manufacturing	478	18
Agriculture[a]	356	13
Petroleum	343	13
Distribution	(200)	(8)
Production[b]	(143)	(5)
Railroads	255	10
Other[c]	500	19
Total	2,652	100

Notes: a. Includes investments in timber. b. Includes refining. c. Sales organisations, utilities, banks and insurance companies.
Sources: C. Lewis, *America's Stake in International Investment* (Brookings Institution, Washington, 1938), p.578ff. and M. Wilkins, *The Emergence of Multinational Enterprise: American Business Abroad from the Colonial Era to 1914* (Harvard University Press, Cambridge, Mass., 1970), p.110.

Table 2.5: Geographical Distribution of US Direct Private Foreign Investment, 1914

Region	Investment ($ millions)	(%)
Western Hemisphere	1,899	72
Canada	(618)	(23)
Mexico	(587)	(22)
Caribbean and Central America	(371)	(14)
South America	(323)	(12)
Eastern Hemisphere	753	28
Europe	(573)	(22)
Asia[a]	(120)	(5)
Other[b]	(60)	(2)
Totals[c]	2,652	100

Notes: a. Excludes investments by Standard Oil (New Jersey) in the Dutch East Indies. b. Africa, Oceania and 'unassigned.' c. Because of rounding the total may not equal the sum of the column.
Sources: C. Lewis, *America's Stake in International Investment* (Brookings Institution, Washington, 1938), p.578ff. and M. Wilkins, *The Emergence of Multinational Enterprise: American Business Abroad from the Colonial Era to 1914* (Harvard University Press, Cambridge, Mass., 1970), p.110.

manufacturing plants in Europe in 1914 were American Radiator, American Tobacco, Diamond Match, Ford, General Electric, Heinz, International Harvester, National Cash Register, Singer Sewing Machine, Western Electric and Westinghouse.[26] All these firms had either a specific technological advantage or a unique product from which they hoped to reap a profit in the European market. They chose the path of direct investment in Europe, rather than exporting from the US, for a number of reasons: high product transportation costs, tariffs in European countries, differences in wage rates and the buttressing of their competitive positions *vis-à-vis* their actual and potential European rivals. Furthermore, advances in technology in communications and transport, like the laying of the trans-Atlantic cable and speedier steamship crossings of the ocean, enabled the top decision-makers of a firm to exercise effective control of operations in geographically dispersed areas, giving a further impetus to the drive to acquire overseas productive facilities.

The other area of investment in manufacturing was Canada. Manufacturing investment there in 1914 was greater than in Europe. In that year 46 per cent of manufacturing investment was in Canada as opposed to 42 per cent in Europe.[27] Manufacturing investment flowed to Canada essentially for the same reasons that it flowed to Europe. But, unlike Europe, Canada was also the recipient of investment in the primary industries. As Table 2.6 shows, 46 per cent of US investments in Canada were in mining, agriculture and petroleum production, while only 2 per cent were in those industries in Europe.

In the rest of the Western Hemisphere investment was largely confined to the primary industries. Investment in Latin America was concentrated in three primary industries (mining, agriculture and petroleum) and in railroads. These four industries accounted for over 85 per cent of US investment in that area. It is interesting to note the differential investment patterns between Latin America, on the one hand, and Europe, and to a lesser extent, Canada, on the other. In that area over which the US exercised hegemony, Latin America, the investments were essentially in the extractive industries. Even the investments in railroads, which were concentrated in Mexico, were related to the extraction of minerals from that country. Contrary to the pattern in Europe and Canada, very little investment went into manufacturing.

The US investment pattern was a typically colonial one. Investments in those areas over which the US had effective political control were

Table 2.6: Distribution of US Direct Private Foreign Investment by Region and by Industry, 1914

	Europe (%)	Canada (%)	Latin America[a] (%)
Manufacturing	35	36	3
Petroleum	24	4	10
Distribution	(23)	—	(2)
Production	(1)	(4)	(8)
Sales Organisations[b]	15	4	3
Utilities	2	1	8
Mining and smelting	1	26	43
Railroads	—	11	14
Agriculture[c]	—	16	19
Other[d]	23	1	1
Totals[e]	100	100	100

Notes: a. Includes Mexico, Central America, the Caribbean countries and South America. b. Excludes petroleum distribution. Includes trading companies and sales branches and subsidiaries of large corporations. c. Includes timber. d. Mainly banks and insurance companies. e. Because of rounding, the total may not equal the sum of the column.
Sources: Based on C. Lewis, *America's Stake in International Investment* (Brookings Institution, Washington, 1938), pp.579, 588; M. Wilkins, *The Emergence of Multinational Enterprise: American Business Abroad from the Colonial Era to 1914* (Harvard University Press, Cambridge, Mass., 1970), p.110.

confined to agriculture and raw materials. In those areas over which the US had no political control, as in Europe, or only partial political control, as in Canada, a substantial portion of its investments went into manufacturing. Thus the US investors looked to Latin America as the supply source for their manufacturing base and to Europe and to Canada as outlets for their investments in manufacturing. Their manufacturing investments in Canada, however, were confined to the already developed regions of that country. In the rest of the country their investments followed the colonial pattern.

By 1914 US firms were playing an increasingly important role in the world economy as a whole and in the Western Hemisphere in particular. The bulk of the US investments were in the extractive industries and in industries directly linked to them, like railroads. Among the extractive industries, traditional mining, like non-ferrous metals, were the most important. By that time however a new industry, oil, had come on to the scene. It was this industry that was destined to play an increasingly important role in the future external expansion of the United States.

Notes

1. *Historical Statistics of the United States, Colonial Times to 1957: A Statistical Abstract Supplement* (US Government Printing Office, Washington, 1960), p.236.
2. The full extent to which the United States was involved in wars against the Indians was revealed in a 1950 publication of the US Armed Forces Information School. This publication lists 99 'wars, expeditions, campaigns, occupations, and other disturbances in which the Army of the United States has participated' in the period between 1775 and 1950. This count includes neither the Korean or Vietnam wars nor separate military activities conducted by the Navy and the Marines. The bulk of these 99 'wars, etc.' took place in the nineteenth century against the Indians. They included such long wars as 'The Apache Indian Wars and Troubles', which lasted 29 years; 'The Navajo Troubles' and the 'Continuous Disturbances with Comanche, Cheyenne, Lipan and Kickapo Indians', each of which lasted 12 years; and the four 'Seminole Wars', which spanned a period of over 80 years. See Armed Forces Information School, *The Army Almanac* (US Government Printing Office, Washington, 1950), pp.439-41.
3. *Historical Statistics of the United States*, p.236.
4. The liquidation of the Indian population was apparently an unwritten objective of the various US governments during this period. This objective was translated into folklore through the popular expression, 'The only good Injun is a dead Injun.' There are no accurate figures on the effectiveness of this genocidal policy. Some estimates place the Indian population of what is now the Continental United States at one million in 1800. The first complete census of Indians, taken in 1890, placed their number at around 250,000. See *Historical Statistics of the United States*, p.9.
5. A trust was a type of combination in which stockholders in competing firms surrendered their stock to a group of trustees in exchange for trust certificates. The trustees were then able to control the competing firms and, in effect, to operate them as one unit. The higher profits resulting from these monopolistic operations were distributed to the owners of the trust certificates as dividends.

The trust was devised by one of Rockefeller's lawyers. Following Rockefeller's initiative in oil, trusts were formed in such industries as whiskey, sugar, lead, linseed oil and several others. The unpopularity of the trusts made them the object of hostile legislation and court decisions. As a result, the trusts were abandoned in favour of the holding company.

The holding company is a device by which one company controls another company through the ownership of stock. Often a holding company can gain control over another company by purchasing only a small share of the latter company's stock. Under legislation passed in the early 1900s, trusts were enabled to become holding companies by transforming the trust certificates into stocks. It was under this legislation that the Standard Oil Trust was transformed into the holding company, Standard Oil of New Jersey.

Each company controlled by a holding company retains its identity. With a merger one company normally buys out the stocks or assets of the other. The bought company usually loses its identity. It is absorbed by the buyer.
6. F.M. Scherer, *Industrial Market Structure and Economic Performance* (Rand McNally, Chicago, 1970), p.48.

7. 'Medway Islands' in *Encyclopaedia Britannica* (24 vols., Encyclopaedia Britannica, Chicago, 1962), vol. 15, p.460.

8. 'Alaska' in *Encyclopaedia Britannica*, vol. 1, p.501.

9. S. Nearing and J. Freeman, *Dollar Diplomacy*, reprint (Monthly Review Press, New York and London, 1966), p.283.

10. *Historical Statistics of the United States*, p.236.

11. Ibid., p.236.

12. Nearing and Freeman, *Dollar Diplomacy*, p.176.

13. Ibid., p.177.

14. The US Constitution bars from the Presidency anyone born outside the United States.

15. W.A. Williams, *The Roots of the Modern American Empire: A Study of the Growth and Shaping of Social Consciousness in a Marketplace Society* (Random House, New York, 1969), p.365.

16. Ibid., p.365.

17. Ibid., pp.440-1.

18. These landings and occupations are too numerous to document in this book. They are described in such works as Nearing and Scott, *Dollar Diplomacy*; Williams, *The Roots of the Modern American Empire*; and M. Wilkins, *The Emergence of the Multinational Enterprise: American Business Abroad from the Colonial Era to 1914* (Harvard University Press, Cambridge, Mass., 1970).

19. 'President Taft Defines Dollar Diplomacy, December, 1912' in W.A. Williams (ed.), *The Shaping of American Diplomacy: Readings and Documents in American Foreign Relations*, 2 vols. (Rand McNally, Chicago, 1956), vol. 2, pp.494-5.

20. C.K. Hobson, *Export of Capital* (Constable, London, 1914), Chs. 1 and 2 and V.I. Lenin, *Imperialism: The Highest Stage of Capitalism* (International, New York, 1969), p.62.

21. *Historical Statistics of the United States*, p.564.

22. Ibid., p.558.

23. Ibid., p.562.

24. C.P. Kindleberger, *American Business Abroad: Six Lectures on Direct Investment* (Yale University Press, New Haven and London, 1969), p.13.

25. *Historical Statistics of the United States*, p.565.

26. M. Wilkins, *The Emergence of Multinational Enterprise*, pp.212-13.

27. Ibid., p.110.

3 THE OIL INDUSTRY'S ROLE IN AMERICAN EXPANSION

The US oil industry came on to the scene in 1859, when 'Colonel' Edwin Drake drilled the nation's first oil well near Titusville, Pa.[1] Though oil was produced and sold commercially in Canada, Russia, Burma and Rumania before 1859,[2] Drake's well was the one that was destined to have the greatest impact on the world. The success of that well was due to a number of factors.

First, Drake, like the typical Schumpeterian innovator, introduced a new technique of production. He drilled for oil instead of digging for it.[3] Drilling reduced significantly the cost of extracting oil, thus making oil competitive with other fuels. Second, technological advances provided a ready market for this oil. Among them were lamps designed to burn oil distilled from coal. These lamps were widely distributed throughout the US, providing a ready outlet and marketing network for the kerosene derived from crude oil. Furthermore, the coal-oil industry, which was the fastest-growing illumination industry in the US during the 1850s,[4] had developed the techniques for refining, which the oil industry adopted.[5] Third, the US was growing at a rapid pace. The outbreak of the Civil War in 1861 greatly accelerated this pace. The expanding industrial base required ever greater quantities of illuminants and lubricating oils, both of which the fledgling oil industry could provide at relatively low prices. Furthermore, this industrialisation was not confined merely to the United States. It was also taking place in Western Europe, which provided an important market for the US oil industry. The US was able to garner this market because its oil fields were far more prolific than those of its closest rival, Canada. In 1862, for instance, the US fields produced more than three million barrels while the Canadian ones produced only 70,000.[6] In fact, in that year all the world's oil was produced in the United States and Canada. The US share was 97 per cent, the Canadian, 3 per cent.[7] Thus this conjuncture of lower costs, new technology, growing industrialisation and prolific fields had the effect of making Drake's well the one that ushered in the oil era.

The Organisation of the Industry

With the growth of this new industry, there also emerged a new group of entrepreneurs, who were eager to control it and propel it forward. These entrepreneurs engaged in both interproduct and intraproduct rivalry. They not only sought markets in areas held by suppliers of other types of fuel, like coal-oil, but in areas held by each other. The very logic of this intense rivalry, as Schumpeter has so aptly pointed out,[8] is for each entrepreneur to strive to obtain monopoly positions, which will both protect him against attacks from others and enable him to attack his rivals more effectively.

This process of intense competition followed by monopolisation became evident in the early years of the oil industry. This process, it should be noted, was independent of the will of the actors. It can be presumed that most, if not all, actors desired to achieve a monopoly position. The ability of any one actor to do so was dependent not only on his individual business acumen but also on the objective conditions in that particular industry. Objective conditions and acumen were the essential ingredients of monopoly power.

Objective Conditions

There were three factors present in the early development of the oil industry which made monopolisation possible. They were: (1) the presence of scale economies in refining; (2) the geographical distance between the major markets and the producing centres; and (3) the limited supplies of crude oil available.

Scale Economies. Scale economies occur when average production costs fall as the size of the production unit increases. The presence of significant scale economies in any industry is incompatible with perfect competition in that industry. If the optimum plant size, i.e. the plant size at which average costs are at a minimum, is large in relation to the size of the market, then that market can support at most only a few plants or, in the extreme case of 'natural monopoly', only one. Scale economies spring from a number of sources, including technology, greater specialisation, and a higher utilisation of scarce factors, like skilled management.

Scale economies appeared in the US refining industry at a relatively early date. Though the data are scanty, the evidence nevertheless indicates that average costs were lower in the larger refineries. In the 1880s, for instance, the average refinery had a capacity of between

1,500 and 2,000 barrels per day. Its average cost was about 2.5¢ per gallon. The costs in the refineries owned by the Standard Oil Trust were stated by the company to be 0.534¢ per gallon.[9] The Trust's refineries were much larger than those of its competitors. The largest one, located at Bayonne, NJ, had a capacity of between 6,000 and 7,000 barrels per day.[10] Though some critics claim that Standard Oil deliberately understated its cost figures to give the impression that its dominant position was due solely to its superior efficiency and though these alleged efficiencies may have been in part due to factors other than scale economies, there are nevertheless indications that scale economies were important. One such indication was the declining number of refineries. Though the demand for products was rising, the number of refineries operating in the US fell from 155 in 1870 to 89 in 1880.[11] This absolute reduction in the face of rising demand could not have occurred unless large-scale units reduced the possible number of competitors, thereby creating conditions for the monopolistic control of the industry.

Geographical Distance. When markets are geographically isolated from sources of supply, transportation costs may become an important factor in determining the specific structure of a given industry. To the extent that transport costs are a significant component of the final price, then a reduction in these costs can give one firm a marked advantage over its competitors. The ability of a firm to achieve advantageous freight rates rests both on its own bargaining position and the structure of the transportation industry. If perfect competition exists in a transportation industry, it would be impossible for one firm to give one shipper more favourable terms than another. Competition will eventually force all firms to charge the same price. If, on the other hand, competition does not exist, then a firm can discriminate among shippers. Such a firm, in fact has an incentive to favour larger shippers over smaller ones. In turn for a guarantee of a constant, high volume of shipments, it can lower rates. Such bilateral arrangements offer many advantages to both parties.

The geographical separation of markets from sources characterised the early US oil industry. The main sources were in Western Pennsylvania and Ohio. The principal markets for refined products were in New England, the Middle Atlantic States and Europe.[12] Furthermore, considerable refining capacity was concentrated in areas like Pittsburgh and Cleveland, which were located between the oil fields and the markets. In 1873, these two cities alone accounted for almost one-half of US refining capacity.[13] Oil going to and from these refining centres thus incurred two transportation charges, the charge on the crude oil flowing from the fields to refineries and the charge on the products going from the refineries to their ultimate destinations. These charges were high relative to the value of the goods. For instance, in the early 1870s the Standard Oil Trust was paying approximately 20 per cent of value for both its crude and product shipments.[14] Because Standard received favourable rates, its competitors had to pay considerably more for transportation.

The chief means of transporting crude were the railroads and pipelines. They were also the chief means of transporting products. In addition, the Cleveland refineries were able to transport products by boat to New York via the Great Lakes and the Erie Canal. Since none of these industries was perfectly competitive, an environment existed which nurtured bilateral arrangements between large shippers of oil and the transportation companies. These arrangements played a major role in shaping the future structure of the oil industry.

Limited Supplies. The existence of a finite stock of a raw material input creates ideal conditions for monopoly control. A monopolist controlling that stock has no fear of new entrants. Furthermore, he can exercise control over the industries using the raw material as an input. Control of this finite stock therefore can give him considerable leverage throughout the economy. The very existence of such a stock thus engenders powerful incentives to control it.

This control can be achieved in a number of ways. One is through contracts and other arrangements for the purchase of raw materials. The other is through buying out the owners of the stock. In the early history of the US oil industry, the former method was used. In later periods, the oil companies attempted to use the latter.

In the years between the drilling of Drake's well and the turn of the century, practically all the oil in the US was located in the fields of Western Pennsylvania and Ohio. Though the ownership of wells in these fields was dispersed among 14,000 producers,[15] the control of production was actually in the hands of a few large refiners, who bought the bulk of crude production. These refiners had little to fear from competition as long as the bulk of the crude came from these fields. These limited supplies satisfied one of the key requirements for monopoly control.

Business Acumen

Objective conditions are a necessary but not sufficient condition to bring about monopoly control. What is also needed is the existence of entrepreneurs who can recognise these conditions and utilise them for their own ends. Presumably the existence of these conditions will, in and of itself, give birth to these entrepreneurs.

Such an entrepreneur was John D. Rockefeller. The role of the entrepreneur, according to Schumpeter, is to create 'the new consumers' goods, the new methods of production or transportation, the new markets, the new forms of industrial organisation'.[16] Because Rockefeller had a better understanding than anyone else of the objective conditions in the oil industry, he was able in a relatively short time to create a 'new form of industrial organisation' by transforming the industry from a competitive one to a monopoly. In doing so he used methods which earned him universal condemnation. Though his severest critics attributed his successes to his questionable practices, they failed to realise that his methods were thoroughly consistent with the mores of competitive capitalism and that they could not have been successful had not the objective conditions for their success been present. If Rockefeller had not changed the organisation of the industry, someone else would have.

A great deal has been written about Rockefeller's tactics. They include such practices as manipulation of freight rates, industrial espionage and the threats against distributors who bought from his competitors. Through promises to guarantee the railroads a minimum quantity of shipments and threats to build pipelines if the railroads did not comply, he obtained rebates on his shipments and drawbacks on the shipments of his competitors.[17]

Although Rockefeller tried to keep his agreements secret, he was often unable to do so. Many were leaked to the public. One such agreement was the South Improvement Company Plan of 1872. Though the Pennsylvania State Legislature declared that Plan, with its highly discriminatory freight rate structure, invalid shortly after its adoption, Rockefeller nevertheless was able to use it to obtain control of the Cleveland refining industry. Ida Tarbell, one of his early biographers, describes how he was able to do so.

> There were at that time some twenty-six refineries in the town —
> some of them very large plants. All of them were feeling more or
> less the discouraging effects of the last three or four years of
> railroad discriminations in favour of the Standard Oil Company. To

the owners of these refineries Mr. Rockefeller now went one by one, and explained the South Improvement Company. 'You see,' he told them, 'this scheme is bound to work. It means an absolute control by us of the oil business. There is no chance for anyone outside. But we are going to give everybody a chance to come in. You are to turn over your refinery to appraisers, and I will give you Standard Oil Company stock or cash, as you prefer, for the value we put upon it. I advise you to take the stock. It will be for your good.' Certain refiners objected. They did not want to sell. They did want to keep and manage their business. Mr. Rockefeller was regretful, but firm. It was useless to resist, he told the hesitating; they would certainly be crushed if they did not accept his offer . . .[18]

Within three months, 21 of the 26 refineries sold out to Rockefeller. Standard Oil's capacity rose from approximately 1,500 barrels per day to 10,000 barrels. By 1872, Rockefeller's Standard Oil Company, which was incorporated only two years earlier, owned more than one-fifth of the nation's refining capacity.[19] The ownership of numerous refineries gave Rockefeller the power to rationalise production by shutting down the small, inefficient units and concentrating it in the larger, more efficient ones. Between 1882 and 1886, for instance, he closed down 31 refineries, confining production to his 22 larger units.[20] This increased efficiency gave Standard an added advantage over its competitors, enabling it to garner a larger share of the market.

The very size of Standard Oil gave it additional power to win further concessions from the railroads. This, in turn, further increased its size and power. By 1899, it owned 82 per cent of US refining capacity.[21] This extensive ownership gave it effective control over crude supplies.

In the years immediately following the discovery of oil in Pennsylvania, crude producers sold their wares to refiners through the Oil City Exchange. This Exchange, functioning like the ideal markets described in textbooks, enabled buyers and sellers to come together. As a result of their transactions a market price was established, which often fluctuated wildly. Perhaps more important than its price role, this Exchange enabled independent crude producers to find markets and independent refiners to find sources of supply.

This particular function became less relevant as the Exchange began to be dominated by one large buyer, Standard Oil. The very size of Standard forced most independent crude producers to sell

to it. Nevertheless some individual producers were able to establish
contacts with independent refiners through the Exchange. This
opportunity for establishing contacts, however, declined in 1895 when
Standard decided to bypass the Exchange and buy directly from
producers in the field at a price posted by itself. This action forced
the Exchange to close,[22] thereby making it extremely difficult for
independent refiners to contact producers.

Another method Rockefeller used to foreclose oil from independent
refiners was to enter into production sharing arrangements. In the late
1880s, when crude prices were falling because of increased production,
he agreed to stabilise prices provided producers reduced their output.
Prices rose as wells were shut-in.[23] The shutting down of wells
decreased the quantities flowing to independent refiners. The rise in
prices encouraged exploration in other Pennsylvania fields, which
Rockefeller sought to control. This control of new oil was
necessary both for replacing oil from the older, depleted wells and
for keeping it out of the reach of independent refiners. While
Rockefeller never did achieve complete control over the Pennsylvania
and Ohio fields and while he was unable to prevent the rise of a
competitor, Pure Oil, he nevertheless managed to gain control of
88 per cent of Pennsylvania and 85 per cent of Ohio crude production
by the turn of the century.[24] Because of this, he was able to restrict
the expansion of rivals such as Pure Oil.

The size of Standard also made most distributors of petroleum
products vulnerable to threats by the company. This vulnerability gave
Rockefeller the power to stop them from buying from his competitors.
To obtain information on distributors, he established an elaborate
network of industrial espionage. Included in this network were
employees of the railroads, who furnished him with data on shipments
by his competitors. Ida Tarbell, who in 1903 came into possession
of reports filed by Standard's agents, described how the network
operated.

They [the documents] show that the Standard Oil Company
receives regularly today, at least from the railroads and steamship
lines represented in these papers, information on *all* oil shipped.
A study of these papers shows beyond question that somebody
having access to the books of the freight offices records regularly
each oil shipment passing the office — the names of consignor and
consignee, the addresses of each, and the quantity and kind of oil
are given in each case . . . Now what is this for? Copies of letters

and telegrams accompanying the reports show that as soon as a
particular report had reached Standard headquarters and it was
known that a car load, or even a barrel, of independent oil was on
its way to a dealer, the Standard agent whose name was written
after the shipment on the record had been notified. 'If you can
stop car going to X, authorise rebate to Z (name of dealer) of
three-quarters cent per gallon,' one of the telegrams reads. There
is plenty of evidence to show how an agent receiving such
information 'stops' the oil. He *persuades* the dealer to countermand
the order.[25]

Both the carrot and the stick were used as methods of 'persuasion'.
Initially the dealer was offered concessions if he countermanded the
order. If he refused, he was threatened with price wars. Tarbell cites
letters submitted to the State of Ohio's 1898 inquiry into the practices
of Standard Oil. A typical letter from a dealer to an independent
distributor or refiner read:

I am sorry to say that a Standard Oil man from your city
followed that oil car and oil to my place, and told me that he
would not let me make a dollar on that oil, and was dogging me
around for two days to buy that oil, and made all kinds of threats
and talked to my people of the house while I was out, and
persuaded me to sell, and I was in a stew what I should do, but I
yielded and I have been sorry for it since. I thought I would hate
to see the bottom knocked out of the prices, but that is why I
did it — the only reason.[26]

Rockefeller's tactics not only enraged the 'little' man but some members
of the 'elite' as well. The Wall Street banker, John P. Morgan, made no
secret of his dislike of Rockefeller.[27] This dislike probably stemmed
more from his business dealings with Rockefeller than from his
disapproval of Rockefeller's tactics. Morgan's own tactics did not, after
all, differ significantly from those of Rockefeller. They were the tactics
of the rising American elite, these Schumpeterian entrepreneurs, who
were out to dominate and control.
    The elite, which had considerable control over the press, constantly
attacked Rockefeller through such papers as Hearst's *New York Journal*
and Pulitzer's *New York World*.[28] They resented this newcomer to their
ranks, who was threatening their dominant position in society. They
were joined by such Populists as US Senator Robert LaFollette of

Wisconsin, who called Rockefeller 'the greatest criminal of the age'.[29]

It was this conjuncture of forces that led the US government to launch its famous anti-trust action against Standard Oil. When it finally ended in 1911, the Supreme Court upheld the decision of the lower court ordering the dissolution of the company. In this decision, which actually served to strengthen monopoly power in the US, the Court found that:

> the very genius for commercial development and organization which it would seem was manifested from the beginning soon begot an intent and purpose to exclude others which was frequently manifested by acts and dealings wholly incompatible with the theory that they were made with the single conception of advancing the development of business power by usual methods, but which, on the contrary, necessarily involved the intent to drive others from the field and to exclude them from the right to trade, and thus accomplish the mastery which was their end in view . . . The exercise of the power which resulted from that organization fortifies the foregoing conclusions, since the developments which came, the acquisition here and there which ensued of every efficient means by which competition could have been asserted, the slow but relentless methods which followed by which means of transportation were absorbed and brought under control, the system of marketing which was adopted by which the country was divided into districts and the trade in each district in oil was turned over to a designated corporation within the combination, and all others excluded, all lead the mind up to a conviction of a purpose and intent which we think is so certain as practically to cause the subject not to be within the domain of reasonable contention.[30]

The Court stressed the term 'reasonable contention' because it found that Standard Oil's actions in restraint of trade to be 'unreasonable'. Presumably, if the actions were 'reasonable', the Court would not have ruled against the company. By adopting this 'rule of reason' the Court gave itself wide powers to determine whether or not a company should be penalised for its behaviour. These powers enabled the Court to take sides in the various disputes among factions of the American elite. Thus, the Court in 1920 used the same 'rule of reason' to absolve the United States Steel Corporation of violations of the Anti-Trust Law, even though the Court admitted that the Corporation engaged in practices which 'may be violations of the law'.[31] The Court found these practices to be 'reasonable' because they were 'transient in their purpose and effect'.[32] The United States Steel Corporation, it should be noted, was established by J.P. Morgan and controlled by Morgan interests. The differential treatment of Rockefeller and Morgan probably was a reflection of power relationships within the American Establishment. The Rockefellers at that time had still not reached the apex of power.

Though the Court dissolved the Standard Oil Company of New Jersey, it did not make any significant dent in Rockefeller's armour. The decision served more to slow his rise than to topple him. It severed Standard's relationship with its subsidiaries by ordering Standard to deliver directly to its shockholders the shares it held in the subsidiaries.[33] In effect this meant that Rockefeller, who held 244,385 shares of Standard's 983,383 shares,[34] received proportionate shares in the 39 constituent companies. For many years after dissolution, the former Standard subsidiaries worked in harmony with each other, though eventually some did strike out on their own. While the decision did stunt Standard's growth, it did not significantly destroy the harmony of interests of the Standard companies, which Rockefeller had so laboriously built up.

In reality, certain events were unfolding, even before the decision, which served to temper the hostility towards Rockefeller. These events flowed from changes in competitive conditions at home and the rise of new companies abroad.

Changes in Domestic Competitive Conditions

Despite all his efforts, Rockefeller was not able to exclude newcomers from the industry. His strategy was to own the refineries and the transportation facilities, leaving the production of crude to others. This strategy was based on the premise that there was only a limited amount of crude, whose production he could control through the ownership of downstream facilities, like refining and distribution. He therefore neglected opportunities to invest in new sources of crude. In 1888, for instance, Standard produced from its own properties less than 200 barrels per day of crude. Total production in the US was around 76,000 barrels per day.[35]

Rockefeller also felt that production of crude was basically unprofitable. He changed his mind when crude prices began to rise in the late 1880s by acquiring producing properties. By 1891 Standard produced more than one-fourth of domestic crude.[36] Nevertheless it did not acquire crude as vigorously as other facilities. Thus, in the

Supreme Court case Standard's attorneys argued that the company was not a monopoly because it controlled 'a very small percentage of the crude oil produced'.[37] Probably the major reason for this 'very small percentage' was Standard's failure to acquire producing properties in Texas.

The US oil picture changed dramatically as a result of the Spindletop discovery in Texas in 1901, one of the great discoveries of all time. Not only did it significantly increase the supply of oil but it also provided the opportunity for new companies to enter. These new entries might never have occurred had Standard ignored the advice of its 'expert'. In 1900 Standard was approached by Captain Anthony Lucas,[38] who was drilling for oil at Spindletop, for financial backing. The company sent Calvin Payne, its authority on oil properties, to survey the terrain. In what was probably one of the most colossal misjudgements in history he reported that there was 'no indication whatever to warrant the expectation of an oil field on the prairies of southeastern Texas' and that Spindletop 'has no analogy to any oil field known, as in fact there was not the slightest trace of even an oil escape'.[39]

There was another reason why Rockefeller refused to invest in Spindletop. The State of Texas, spurred on by one of its leading politicians, James Hogg, had launched legal actions against the Rockefeller companies operating within its boundaries. It even passed a law which effectively prohibited Rockefeller from doing any business in the State.[40] Rockefeller was determined to 'punish' Texas by refusing to invest in it. Thus, even after Spindletop was proven to be a success, Standard turned down a chance to buy into it. Representatives of the company reportedly told the Mellon family of Pittsburgh, who approached Standard to sell their investment in Spindletop, that 'after the way Mr. Rockefeller has been treated by the State of Texas, he'll never put another dime in Texas'.[41]

As it turned out, it was Rockefeller, rather than Texas, who was 'punished' by this decision. The Mellons, who financed Spindletop after Rockefeller initially refused to do so, stayed in the oil business and organised Gulf Oil, which became a major competitor of Standard. In addition, Texas financial interests brought together by Jim Hogg established the Texas Company, which also threatened Standard's supremacy. The possession by both of these companies of large reserves of Texas oil gave them a distinct advantage over Standard with its relatively small crude supplies.

By the time of its dissolution in 1911 Standard controlled only 10 per cent of oil in the fields of the Texas Gulf Coast. Its control of refinery capacity had fallen to 64 per cent from a high of approximately 95 per cent 30 years earlier.[42] The new entrants, whose power was based on the ownership of crude rather than refining capacity, were able to breach Rockefeller's domestic monopoly even before the Supreme Court acted. Furthermore, new entrants abroad, whose power was also based on the possession of crude, posed an additional threat to Rockefeller's dominance of foreign markets.

New Companies Abroad

In the latter part of the nineteenth century, two important new oil centres arose in Russia and in Indonesia. Russian oil was controlled by the Nobel and the French Rothschild interests while Indonesian oil was controlled by Dutch and English interests. These groups competed vigorously among themselves and with Rockefeller for control over the world's supplies and markets. In this struggle, each of these groups was backed by its home government. Thus, the US government, while ostensibly punishing Rockefeller for his actions at home, gave considerable support to his activities abroad. This support was an essential ingredient of its policy of expanding exports.

Foreign Trade

From its very beginning oil was important in America's foreign trade. In 1866, only seven years after the Drake well and only three years after John D. Rockefeller entered the oil business, John sent his brother, William, to New York to set up an export office.[43] By the time Rockefeller had established this office, exports of refined products exceeded domestic sales. In 1866, for instance, more than two-thirds of refinery output was exported. By 1873 exports accounted for almost three-fourths of refinery output.[44] This high portion of exports continued almost to the turn of the century, when two developments, the rise of competition abroad and the growth of the domestic market, reduced it.

Another aspect of the foreign trade in oil is that it consisted mainly of trade in refined products rather than in crude. This pattern was strikingly different from that exhibited in the rest of America's foreign trade. By the end of the US Civil War American industrialisation lagged behind European. The US continued to play its historic role as exporter of raw materials and foodstuffs and importer of finished goods. As can

be seen in Table 3.1, US exports of finished goods did not exceed those
of raw materials until the 1890s. In fact, 1893 was the first year in
which the value of finished goods exports was greater than that of
unfinished goods.[45]

Table 3.1: Composition of US Exports, 1866-1914 (per cent)

Years	Primary goods	Finished goods[a]
1911-14	39	61
1906-10	41	59
1901-05	42	58
1896-1900	45	55
1891-5	51	49
1886-90	53	47
1881-5	55	45
1876-80	56	44
1871-5	60	40
1866-70	67	33

Note: a. Includes semi manufactures.
Calculated from: *Historical Statistics of the United States, Colonial Times to 1957: A Statistical Abstract Supplement* (US Government Printing Office, Washington, 1960), pp.544-5.

Refined oil, on the other hand, became a leading export item long
before 1893. For many years it was the second most important
manufactured export, exceeded only by meat products.[46] As Table 3.2
shows, petroleum products accounted for a significant share of
finished goods exports in the pre-World War I years. This share ranged
from a high of 15 per cent in the late 1880s to a low of 10 per cent in
the early 1900s.

This trade in oil naturally aroused the interest of US government
leaders, who were intent on defending and expanding export markets.
US diplomatic representatives had played an important role in
promoting sales of US oil to foreign countries since the early 1860s.[47]
One of the most overt moves backing oil exports was the declaration
in the late 1890s of the 'Open Door' policy towards China by US
Secretary of State, John Hay. While Hay enunciated this policy,
designed to open up China's markets to US interests, mainly in response
to pressure from exporters of cotton and from financial groups wanting
to build railroads in China,[48] there is little doubt that he was also
interested in expanding US oil markets there. As can be seen in

Table 3.2: Share of Petroleum Products in US Finished Goods Exports, 1886-1914 (per cent)

Years	Petroleum products
1911-14	10
1906-10	10
1901-5	10
1896-1900	11
1891-5	11
1886-90	15

Calculated from: *Historical Statistics of the United States, Colonial Times to 1957: A Statistical Abstract Supplement* (US Government Printing Office, Washington, 1960), pp.544, 546.

Table 3.3: Geographical Distribution of US Exports of Illuminating Oil, 1874 and 1899 (per cent)

Geographical Area	1874	1899	Per cent change, 1874-99
Europe	87	63	−20
Asia	4	20	473
(China)	(1)	(4)	(814)
South America	3	7	187
North America	3	3	25
Other	3	7	70
Totals	100	100	9

Calculated from: H. Williamson and A. Daum, *The American Petroleum Industry, The Age of Illumination* (Northwestern University Press, Evanston, Ill., 1959), pp.742-3, 752.

Table 3.3, the Asian market, as a whole, and the Chinese market, in particular, were becoming increasingly important to the US oil industry. These markets were replacing the losses which the industry was suffering in Europe following the rise of Russian oil. By the turn of the century Asia and China were the fastest growing markets for American oil. These growing markets were however being threatened by the French Rothschilds, who were marketing Sumatran oil there.[49] The 'Open Door' policy unquestionably served to buttress the US oil position in China *vis-à-vis* its competition.

The prime beneficiary of US oil policy was, of course, Rockefeller, who had a virtual monopoly in the export trade. His only significant US competitor was Pure Oil Company, which had only 5 per cent of the market.[50] The benefits of the 'Open Door' policy to Rockefeller were quite substantial. Standard Oil's share of the Asian market in illuminating oil rose from 34 per cent in 1899 to 46 per cent in 1908.[51] Rockefeller, in his *Random Reminiscences*, openly acknowledged the aid rendered him by the US government. 'One of our greatest helpers', he wrote, 'has been the State Department in Washington. Our ambassadors and ministers and consuls have aided to push our way into new markets to the utmost corners of the world.'[52] This close link between the US State Department and the Rockefeller-controlled oil companies has, as we shall see, continued to the present day. What is especially interesting is that this link remained firm even in those periods when the US government was prosecuting these companies for their domestic activities.

Foreign Investment

This aid from the US government was important to Rockefeller in meeting his competition abroad where, unlike in the US, he encountered major competitors with enormous reserves of crude. Rockefeller tried to counter this competition through the acquisition of refining and distribution facilities abroad. Thus, at a relatively early stage in its history, Standard Oil began to make direct investments overseas.

It is interesting to note that Standard initially made market-oriented, rather than supply-oriented, investments. This followed the pattern in the United States, a pattern that enabled him to achieve a dominant position in the domestic market. Furthermore, Rockefeller, believing he could triumph over his foreign competitors, originally eschewed making any market-sharing arrangements with them. His subsequent failure to destroy his competitors forced him both to acquire producing properties and to enter into cartel agreements.

The first major foreign investment by Standard Oil was the establishment in 1888 of the Anglo-American Oil Co to market US refined products in England.[53] Several years later he formed the Deutsch-Amerikanische Petroleum Gessellschaft (DAPG) to perform the same function in Germany.[54] Rockefeller used DAPG not only to pre-empt the German market from Russian oil but also to pre-empt it from his US competitors. In 1896, six years after its incorporation, DAPG bought out Philip Poth, its main competitor in Germany.[55]

Poth had been distributing oil supplied by the Pure Oil Company.[56]

Around the same time Standard set up a French affiliate, Bedford et Compagnie, which purchased an existing refinery and began construction of a new one. This move aroused strong opposition within the French refining industry. The dispute was resolved through a cartel agreement between Rockefeller and the French refiners. The accord provided that all French companies buy the bulk of their crude from Standard. In return Standard agreed not to sell most of its refined products to anyone outside the French group.[57] Thus Rockefeller began using in Europe the same exclusionary tactics which made him so powerful in the United States.

Standard at various times either purchased or built refineries in Cuba,[58] Puerto Rico, when it was still under Spanish rule,[59] Germany,[60] Austro-Hungary, Japan and Canada.[61] The acquisition of refineries in Canada came about through the purchase of controlling interest in Imperial Oil, Canada's largest oil company, in 1898. Prior to this purchase, Imperial, which was incorporated in 1880, gave Standard vigorous competition in the Canadian market. Through this stock acquisition, Standard not only eliminated Imperial as a competitor but also obtained a fully-integrated company which had producing properties as well as refining, transportation and marketing facilities.[62]

Through control of imperial, Standard achieved a dominant position in the Canadian market. It also used Imperial as a vehicle for investing in those countries where Americans were not welcome. When, on the eve of World War I, Standard decided to acquire producing properties abroad, it used Imperial to obtain the controversial La Brea y Parinas fields in Peru. Through Imperial, Standard organised the International Petroleum Company (IPC), ostensibly a Canadian company, to run its fields and refineries in that country. IPC subsequently became the leading oil company in Peru.[63]

Standard used another affiliate, Waters-Pierce Oil Company, to penetrate the Mexican market. Waters-Pierce, Standard's exclusive marketing firm in the US South-West,[64] built three refineries in Mexico, which purchased all their crude from Standard,[65] and acquired producing properties in that country.[66] Standard also used its Dutch marketing affiliate, the American Petroleum Company, to acquire concessions in Indonesia in 1912.[67] It used the affiliate route of acquisition because in 1899 the Dutch government blocked its direct attempts to obtain concessions there.[68]

The British government also stopped it from investing in Burma's oil fields in 1902.[69] Standard did not attempt to circumvent this prohibition, Rumania, on the other hand, had no objection to Standard acquiring producing properties within its boundaries. Standard acquired 100 per cent ownership of Romano-American Company, a fully-integrated company with both refineries and oil fields.[70]

All told, before its dissolution in 1911, Standard had 67 affiliates engaged in foreign trade.[71] By the end of 1907, it controlled 55 foreign companies with a capitalisation of approximately $37 million.[72] Nevertheless Rockefeller was never able to achieve the dominance in foreign markets that he attained in the domestic one. In 1910, one year before dissolution, Standard Oil's share of overseas crude production was less than 1 per cent.[73] As far as products were concerned, Standard's share of the overseas petroleum market never reached one-third in the years before World War I. As indicated in Table 3.4, the share of all US companies in the foreign market fell from approximately one-fourth to one-fifth in the early years of the century and then rose to approximately 30 per cent by 1914.

In the years before the war several other US companies entered the foreign field. They included Pure Oil, which established its own marketing facilities in Europe after DAPG acquired Philip Poth.[74] These however were subsequently sold to Shell and Standard.[75] The Texas Company began investing in sales outlets throughout the world. It also invested in producing properties in Mexico, as did Gulf, Sinclair and a host of other American companies.[76]

This overall growth of direct foreign investments by the petroleum industry was, as was indicated in chapter 2, part and parcel of the general overseas expansion of American business in the pre-World War I era. There was nothing particularly spectacular about the growth of the oil industry beyond the US borders. Its growth was, in fact, slower than the growth of total direct foreign investment, as can be seen in Table 3.5.

There were several reasons for this relatively slow growth. One, the growing domestic market for oil, especially for use in automobiles, tended to keep investment at home. Two, many foreign governments discouraged the entry of Rockefeller-owned companies in their countries. Three, Rockefeller's policy of investing in refining and marketing rather than in extraction restricted investments to those countries where living standards were relatively high. Fourth, and probably most important, the US government made no special effort to promote oil investment abroad. Though the government did encourage and help oil companies invest overseas, it did so as part of a

Table 3.4: US Share of World Petroleum Market Outside the United States, 1899-1914 (per cent)

Year	Share
1914	31
1913	30
1912	27
1911	26
1910	24
1909	25
1908	27
1907	25
1906	27
1905	27
1904	19
1903	19
1902	21
1901	21
1900	21
1899	23

Calculated from: United States Senate, *Petroleum Interests in Foreign Countries: Hearings before a Special Committee Investigating Petroleum Resources* (US Government Printing Office, Washington, 1946), p.214.

Table 3.5: Per Cent Changes in US Direct Foreign Investment by Industry, 1897-1914

Industry	Change
Other[a]	796
Utilities	504
Mining and smelting	437
Manufacturing	408
Agriculture and timber	362
Petroleum[b]	304
Sales organisation[c]	198
Railroads[d]	77
Total	318

Notes: a. Mainly banks and insurance companies. b. Includes exploration, production, refining and distribution. Excludes investments in Indonesia. c. Includes trading companies and sales branches and subsidiaries of large corporations. Excludes petroleum distribution. d. Includes US government purchase of Panama Canal Railroad.
Calculated from: C. Lewis, *America's Stake in International Investment* (The Brookings Institution, Washington, 1938), pp.579, 588; M. Wilkins, *The Emergence of Multinational Enterprise, American Business Abroad from the Colonial Era to 1914* (Harvard University Press, Cambridge, 1970), p.110.

general policy of spreading American business abroad. It saw no reason why it should give preferential treatment to the oil companies.

The events of World War I dramatically changed the US government's attitude. That war demonstrated the importance of oil in modern warfare. From that time on, obtaining control of oil became one of the cornerstones of US foreign economic policy. It became the essential ingredient in the expansion and consolidation of the American Empire.

Notes

1. 'Colonel' Drake was in reality a railroad conductor. The New England promoters backing his project addressed letters to him with the title of Colonel in order to impress the residents of Titusville. This type of promotional activity was not untypical of future oil industry behaviour. See H.F. Williamson and A.R. Daum, *The American Petroleum Industry, The Age of Illumination* (Northwestern University Press, Evanston, Ill., 1959), pp.75-6.

2. H. O'Connor, *World Crisis in Oil* (Monthly Review Press, New York, 1962), p.27.

3. In 1857, two years before the successful completion of Drake's well, James Williams, a Canadian businessman, used a combination of digging and drilling to obtain oil in Ontario. Drake, however, concentrated entirely on drilling. See E. Gray, *Impact of Oil, The Development of Canada's Oil Resources* (The Ryerson Press/Maclean Hunter, Toronto, 1969), pp.5-6.

4. Williamson and Daum, *The American Petroleum Industry*, p.40.

5. One of the leaders in developing the refining process was Dr Abraham Gesner, a Canadian physician. See ibid., pp.44-8.

6. Gray, *Impact of Oil*, p.7.

7. Ibid., p.8.

8. J. Schumpeter, *Capitalism, Socialism, and Democracy*, 2nd edn (Harper and Brothers, New York, 1947), Ch. 7.

9. Williamson and Daum, *The American Petroleum Industry*, pp.483-4.

10. Ibid., p.483.

11. Ibid., p.471.

12. Ibid., pp.320, 325.

13. Ibid., p.473.

14. Calculated from ibid., pp.326, 349, 375.

15. Ibid., p.567.

16. Schumpeter, *Capitalism, Socialism and Democracy*, p.83.

17. A rebate is a refund to the shipper on his own shipments. A drawback is a payment to one shipper on the shipments of others. Thus the payments Rockefeller's competitors made to the railroads included not only the regular shipping charges but payments to Rockefeller as well.

18. I. Tarbell, *The History of the Standard Oil Company* (2 vols., The Macmillan Company, New York, 1933), vol. 1, p.63.

19. Ibid., pp.67-8.

20. Williamson and Daum, *The American Petroleum Industry*, p.474.

21. H. Williamson, R. Andreano, A Daum and G. Klose, *The American Petroleum Industry, The Age of Energy, 1899-1959* (Northwestern University Press, Evanston, Ill., 1963), p.7.

22. G. Montague, *The Rise and Progress of the Standard Oil Company*

(Harper and Brothers, New York, 1903), p.131.
    23. Williamson and Daum, *The American Petroleum Industry*, pp.564-8.
    24. Williamson, Andreano, Daum and Klose, *The American Petroleum Industry*, p.7.
    25. I. Tarbell, *The History of The Standard Oil Company*, vol. 2, pp.40-1.
    26. Ibid., p.43.
    27. P. Collier and D. Horowitz, *The Rockefellers: An American Dynasty* (New American Library, New York, 1976), p.54.
    28. R. O'Connor, *The Oil Barons: Men of Greed and Grandeur* (Little, Brown and Company, Boston, 1971), p.59.
    29. Collier and Horowitz, *The Rockefellers*, p.4.
    30. Standard Oil Co. of New Jersey v. United States, 221 US 1 (1911).
    31. United States v. United States Steel Corporation *et al.*, 251 US 417 (1920).
    32. Ibid.
    33. Standard Oil Co. of New Jersey v. United States.
    34. Collier and Horowitz, *The Rockefellers*, p.58.
    35. Williamson and Daum, *The American Petroleum Industry*, p.605.
    36. Ibid., p.607.
    37. Standard Oil Company of New Jersey v. United States.
    38. Lucas was the Anglicised version of Luchich, his family name in the place of his birth, which was then the province of Dalmatia in the Austro-Hungarian Empire and is now Yugoslavia. He preferred the Anglicised version. See H. O'Connor, *The Empire of Oil* (Monthly Review Press, New York, 1962), p.8.
    39. R. O'Connor, *The Oil Barons*, p.71.
    40. C. Solberg, *Oil Power* (Mason/Charter, New York, 1976), pp.53-4.
    41. Ibid., p.59.
    42. Williamson, Andreano, Daum and Klose, *The American Petroleum Industry*, p.7.
    43. Williamson and Daum, *The American Petroleum Industry*, p.302.
    44. Calculated from ibid., pp.737-8.
    45. *Historical Statistics of the United States, Colonial Times to 1957: A Statistical Abstract Supplement* (US Government Printing Office, Washington, 1960), p.544.
    46. Ibid., p.546.
    47. Williamson and Daum, *The American Petroleum Industry*, pp.324, 493.
    48. C.S. Campbell, Jr, 'American Business Interests and the Open Door in China,' *The Far Eastern Quarterly*, vol. 1, no. 1 (1941), pp.43-4.
    49. Williamson, Andreano, Daum and Klose, *The American Petroleum Industry*, p.258.
    50. Ibid., p.251.
    51. Ibid., p.257.
    52. Collier and Horowitz, *The Rockefellers*, p.40.
    53. M. Wilkins, *The Emergence of Multinational Enterprise: American Business Abroad from the Colonial Era to 1914* (Harvard University Press, Cambridge, 1970), p.64.
    54. Williamson and Daum, *The American Petroleum Industry*, p.648.
    55. Ibid., p.653.
    56. Ibid., p.569.
    57. Ibid., p.652.
    58. Ibid., p.661.
    59. Ibid.
    60. C. Lewis, *America's Stake in International Investment* (The Brookings Institution, Washington, 1938), p.581.

61. Ibid., p.582.
62. Wilkins, *The Emergence of Multinational Enterprise*, p.140.
63. Ibid., p.186.
64. Williamson and Daum, *The American Petroleum Industry*, p.543.
65. Lewis, *America's Stake in International Investments*, p.589.
66. Wilkins, *The Emergence of Multinational Enterprise*, p.85.
67. Ibid., pp.85-6.
68. Ibid., p.84.
69. Ibid.
70. Ibid.
71. Ibid.
72. Ibid., p.83.
73. Williamson, Andreano, Daum and Klose, *The American Petroleum Industry*, p.258.
74. Williamson and Daum, *The American Petroleum Industry*, p.577.
75. Wilkins, *The Emergence of Multinational Enterprise*, p.86.
76. Ibid.; Lewis, *America's Stake in International Investments*, p.589.

4 THE RELATIONSHIP BETWEEN THE OIL INDUSTRY AND THE AMERICAN STATE — THE INTERWAR PERIOD

In actuality the British, rather than the Americans, were the first to recognise the military potential of oil. As early as 1899, Marcus Samuel, the head of Shell, began campaigning to convert the British Navy from coal to oil.[1] Initially the admirals resisted Samuel's suggestion, suspecting that it had been motivated by self-interest. They also had doubts about the technical feasibility of using oil.

Churchill's Conversion to Oil

Nevertheless Samuel managed to convince the First Sea Lord, Admiral Fisher, of the soundness of his ideas. Whether Samuel's offer of a Shell directorship to Fisher played a role in his conversion is still a matter of conjecture.[2] In any event Fisher waged a vigorous campaign to transform the Navy into an oil-burning fleet. He found an ally in Winston Churchill, who became First Lord of the Admiralty in 1911.

Fearing the growing strength of the German Navy, Churchill wanted to ensure the supremacy of the British fleet in the war that he knew was sure to come. In 1912 he established The Royal Commission on Oil Supplies, headed by Lord Fisher, to examine the advantages of oil.[3] Not surprisingly the Commission came out in favour of oil. This decision turned out to be a sound one as it did give the British Navy a distinct advantage over the German Navy in World War I. In later years Churchill outlined the superiority of oil.

> In equal ships, oil gave a large excess of speed over coal. It enabled their speed to be attained with far greater rapidity. It gave 40 percent greater radius of action for the same rate of coal. It enabled the fleet to refuel at sea with facility . . . The use of oil made it possible in every type of vessel to have more gun-power and more speed for less size and less cost.[4]

Though the Royal Commission showed that the technical objections to oil were groundless, it still had to deal with the problem of security of supply. Britain had ample coal but practically no oil. Unless secure oil supplies were obtained, the British fleet could be rendered inoperative in a conflict.

The most obvious supplier of oil was Shell. Marcus Samuel, a solid member of the British Establishment, had, after all, initiated the idea. But there was strong opposition to Shell among members of the Commission and in Parliament. Shell had increased the price of oil, arousing a great deal of popular resentment. Sir Marcus (he was knighted in the early 1900s[5]) defended the increase, stating that 'the price of an article is exactly what it will fetch'.[6] This attitude caused Churchill to remark:

> It is their policy . . . to acquire control of the sources and means of supply, and then to regulate the production and the market price . . . We have no quarrel with Shell. We have always found them courteous, considerate, ready to oblige, anxious to serve the Admiralty and to promote the British Navy and the British Empire — at a price.[7]

But perhaps more fundamental was the fear that Shell was not under British control. Several years earlier Rockefeller attempted to buy out Shell. Though Samuel rejected this offer,[8] he merged his company with Henri Deterding's Royal Dutch Petroleum to form Royal Dutch Shell.[9] Under the terms of this 1906 agreement, the Dutch interests obtained 60 per cent of the shares with British interests holding the remainder.

Churchill actually regarded Shell as a Dutch company and suspected that Deterding was under German influence.[10] He did not want such a company to be the main supplier for the British Navy. As it turned out, Churchill's fears were ill-founded. During the war the company, despite the proclaimed neutrality of Holland, sided with Britain. It did not overcharge the British Navy, even though it was in an excellent position to do so.[11]

Another objection was that none of Shell's producing properties was in the British Empire. They lay in Borneo and in the Dutch East Indies. The British could not be sure that these supplies would be available to them in a war. They wanted to be supplied by a company with reserves within the Empire. The only such company in existence was Burmah Oil, which had been producing oil in Burma since 1886.[12] Burma, at that time, was in the Empire. That company furthermore was entirely controlled by British interests.

Burmah's real wealth however lay not in Burma but in Persia. In 1905, Burmah, at the instigation of the British Government, helped finance William D'Arcy, a British engineer, who in 1901 obtained a concession covering 500,000 square miles from the Persian Government. Unsuccessful in his early attempts to find oil, D'Arcy was on the verge of bankruptcy when the British Government asked Burmah Oil to bail him out. In 1905 Burmah founded the Concessions Syndicate to handle its Persian venture. When oil was finally discovered in 1908, Concessions Syndicate formed the Anglo-Persian Oil Company.[13] This company changed its name to Anglo-Iranian in 1935 and to British Petroleum in 1954, the name by which it is presently known.[14]

Though technically this Persian oil was outside the British Empire, it was well within the British sphere of influence and could be defended from bases in India. In fact the British had already sent a gunboat and a detachment of Bengal Lancers to Persia to protect the company's properties against attacks by marauding tribesmen.[15] Thus, the Anglo-Persian Oil Company, controlled by British interests and with reserves within the British sphere of influence, was looked upon as the ideal company to supply the Royal Navy.

By 1913, when the Royal Commission concluded its hearings, this company was having financial difficulties. Though it had invested large sums in Persia, it still had not sold a single barrel of oil.[16] There was a danger that this troubled company might fall into the hands of foreign interests, especially those of Rockefeller's Standard Oil. Churchill was determined that this should not happen. He therefore proposed to a somewhat shocked Parliament that the British Government buy a controlling interest in that company to ensure that it would always remain in British hands.

> We must become the owners, or at any rate the controllers at the source, of at least a proportion of the supply of natural oil which we require . . . and obtain our oil supply, so far as possible, from sources under British control, or British influence.[17]

After considerable debate, intermingled with accusations of 'socialism', Parliament in 1914 approved the purchase by an over-whelming vote of 254 to 18.[18] The government paid £2 million for a 51 per cent share in the company. In return it received a stipulation that Anglo-Persian must always remain a British concern and that every director must be a British subject. Though the government acquired the majority of shares, it obtained the right to appoint only two of the company's seven directors. Furthermore, the government assured the companies that these directors, who had the right of veto, would exercise that right only in matters involving foreign or military policy or Admiralty contracts.[19] This is one promise that the British kept. Neither that government nor any subsequent one has ever exercised the right to veto.[20] In other words, the British permitted private interests to have free rein with public funds.

This marriage of private and public interests marked a new era in the development of British capitalism. As Hartshorn has remarked:

What is germane in this context is that the decision to buy control of this company represented the first open manifestation of the vital interest to the government of a country lacking indigenous oil supplies of securing a source of supply that it recognized as strategically vital. But it was the prototype. Britain sought and gained command of secured 'tied' supplies of oil from a promising source that later turned out to be part of the richest oil-bearing region yet discovered. Over the years it extended that command in the region . . .[21]

Oil Imperialism

What Hartshorn described is the policy of oil imperialism adopted by Churchill, i.e. a deliberate policy of acquiring oil reserves in foreign lands. No longer was Britain, the traditional bastion of free trade, content to rely on the market-place to provide it with this vital commodity. In its quest for oil it abandoned *laisser-faire* capitalism in favour of a state-owned company buttressed by the full might of the Empire.

It was inevitable that other nations would follow Britain's example. World War I underscored the strategic importance of oil. Not only was oil used to power the ships but it was also used in the new instruments of warfare: the tank, the airplane and the lorry. French President Clemenceau recognised this new reality when he wrote US President Wilson that oil was 'as necessary as blood'.[22] He was echoed by his oil commissioner, Henri Berenguer, who proclaimed that he 'who has oil has Empire' and that oil 'is more precious, more penetrating, more influential in the world than gold itself'.[23] Not surprisingly,

after the war ended the French Government followed a policy similar
to England's. It bought shares in CFP (*Compagnie Française des
Pétroles*) in order to guarantee for itself a portion of that company's
reserves in Iraq.[24]

The war also brought home the significance of oil to the United
States. That country furnished 80 per cent of the Allies' petroleum
needs.[25] In order to do so and to supply its armed forces as well, it
imported oil from Mexico[26] and instituted a programme of 'gasless
Sundays' in which people voluntarily refrained from using their
automobiles except for 'errands of necessity'.[27]

These war-time shortages gave rise to fears that the United States
might not have sufficient oil reserves to see it through another war.
Oil thus began to be a matter of special concern to US policy makers.
Mark Requa, director of the oil division of the US Fuel Administration
advocated a policy of encouraging 'in every way possible . . . the
acquisitions by our nationals of reserves in foreign lands'.[28] These
foreign reserves were to be considered part of the US oil supply.

It is interesting to note that US oil policy differed somewhat
from the policies of England and France. The latter countries used
companies in which the governments had invested while the United
States relied wholly on privately-owned companies to achieve the same
end – the attainment of a secure source of oil.[29] This difference
stemmed from the changes in international investment wrought by the
war. At the beginning of the war in 1914, the US was a net debtor,
owing the rest of the world $3.7 billion. By 1919 it had shifted to a net
creditor position with claims on the rest of the world of $3.7 billion.[30]
The US had become the main supplier of capital to the world. There
was little danger that any large American firms would be bought out by
foreigners and thus there was no need for the government to assume
an equity position in any company to keep it under domestic control.
Besides, by the end of the war there were a large number of US oil
firms in existence. If the government wanted to ensure domestic
control through assuming equity positions, it would either have to
buy control of all of them or select only a limited number of them
for this purpose.

The first alternative would have been quite costly. In addition it
would have been viewed as a step towards the nationalisation of the oil
industry, a step which would have been politically unacceptable. The
second alternative would have involved the government in the politically
embarrassing choice of selecting which companies to purchase.
Furthermore, it would have placed these government-controlled

companies in competition with the privately-controlled ones, again
leading to political difficulties. Thus, the government, instead of
selecting a single or small group of companies to use as its agent,
assigned that role to all private oil companies operating abroad. By the
end of the war US policy shifted from one of encouraging foreign
operations of US oil companies as part of a general policy of aiding
the overseas expansion of all US firms to one of giving top priority to
oil. Oil henceforth was to receive special treatment in US foreign
economic policy. The US had formally entered the era of oil
imperialism.

In pursuing its objectives the US adopted two distinct and
contradictory doctrines, those of 'minimum duty' and the 'open door'.
It used the former in its dispute with Mexico and the latter in its
controversies with England and Holland.

Minimum Duty

The doctrine of 'minimum duty' emerged in the argument with Mexico
over that country's attempt in 1918 to nationalise sub-surface mineral
rights. During the latter part of the nineteenth century Porfirio Diaz,
the President of Mexico, granted large concessions to railroads, mines
and agricultural enterprises. These concessions included ownership of
sub-surface resources. Since in most parts of the world outside the
United States these resources were considered the property of the state,
many Mexicans questioned the legality of these grants.

Among those receiving such a concession was Edward L. Doheny,
an American. In 1901 Doheny discovered oil on his land and Mexico
entered the oil era. Oil companies from the United States and England
swarmed into Mexico seeking concessions from the government.
Behind the scenes a bitter struggle ensued between the US and British
Governments over control of Mexico. Diaz, who did not want to be
dominated by any single power, welcomed the British as a counter-
weight to the Americans. Diaz's pro-British policies caused concern
among American investors in Mexico. Shortly after he was elected
President for the eighth time in 1910 he was ousted in a revolution
led by Francisco Madero. Subsequent testimony before the US
Senate Foreign Relations Committee indicated that Madero received
funds from 'several leading bankers in El Paso' and from Rockefeller's
Standard Oil.[31] There was no evidence of direct US government
involvement in the ousting. Nevertheless, Nearing and Freeman noted
that 'the sentiment in the United States was strongly in favour of
Madero and the United States Government took the earliest possible

opportunity of recognizing him'.[32]

The United States, on the other hand, did not extend recognition to Victoriano Huerta, who overthrew Madero in 1913. Huerta was financed by English interests.[33] Britain's support of Huerta infuriated the US Government. Lawrence, in his biography of President Wilson, stated:

> Mr. Wilson had many tempestuous conflicts with the British Foreign Office over the apparent support given to the Huerta regime by Sir Lionel Carden, the British Minister to Mexico, a support intensified to no small extent by the large British oil companies in Mexico whose influence in London circles was appreciable.[34]

Wilson regarded this British support as an obstacle to his attempts to overthrow Huerta. After much wrangling Wilson induced the British to withdraw their support. In return the US agreed to abolish tolls on shipments through the Panama Canal.[35] After this agreement the US Government moved against the Huerta regime. It openly sold arms to Venustiano Carranza and Francisco Villa, who were leading a rebellion against the government. The US also sent troops into the port city of Vera Cruz to prevent the Huerta forces from receiving arms and to deprive it of much-needed customs revenues.[36] Through these tactics the US forced Huerta from office and installed Carranza as President on the eve of World War I.[37]

Carranza's forces also received financial support from private US oil interests, like Doheny[38] and Pierce,[39] of the Rockefeller-controlled Waters-Pierce Oil Company. They had good reason to back a political leader who presumably favoured their interests. These and other American interests had a huge stake in Mexican oil. As can be seen in Table 4.1, US companies had investments of $85 million in production facilities in 1914. Almost three-fifths of oil production investments were in Mexico. Mexico had by that year become a major producer. Its oil output of 26 million barrels was exceeded only by that of the United States and Russia.[40] Approximately two-thirds of this quantity was exported to the United States.[41]

In the years following Carranza's accession to power US oil investments in Mexico, Mexican production and Mexican exports to the US continued to grow. By 1919, as Table 4.1 shows, investments had risen to $200 million. The US firms in Mexico, in addition to those controlled by Doheny and Pierce, were Gulf, Sinclair, Atlantic and Standard Oil (N.J.).[42] As Table 4.2 indicates, Mexican production

Table 4.1: Geographical Distribution of US Direct Foreign Investment in Oil Production,[a] 1914 and 1919

	1914 Investment ($ millions)	(%)	1919 Investment ($ millions)	(%)
Western Hemisphere	135	94[b]	321	98
Mexico	(85)	(59)	(200)	(61)
Canada	(25)	(17)	(20)	(9)
Peru	(15)	(10)	(45)	(14)
Venezuela	(5)	(3)	(18)	(5)
Trinidad	(3)	(2)	(5)	(2)
Colombia	(2)	(1)	(20)	(6)
Central America	(0)	(0)	(3)	(1)
Eastern Hemisphere	8	6	7.5	2
Rumania	(5)	(3)	(7.5)	(2)
Russia	(3)	(2)	(0)	(0)
Totals	143	100	328.5	100

Notes: a. Includes refining. b. Because of rounding the total may not equal the sum of the column.
Source: C. Lewis, *America's Stake in International Investment* (Brookings Institution, Washington, 1938), p.588.

Table 4.2: World Production of Crude Oil, 1914 and 1919 (millions of barrels)

	1914 Production	(%)	1919 Production	(%)	Per cent change 1914-1919
World	407.5	100	555.9	100	36
Major producing areas	392.5	96	535.4	96[a]	36
United States	(265.8)	(65)	(378.4)	(68)	42
Russia	(67.0)	(16)	(31.7)	(6)	−53
Mexico	(26.2)	(6)	(87.1)	(16)	232
Rumania	(12.8)	(3)	(6.6)	(1)	−48
Dutch East Indies	(11.4)	(3)	(15.5)	(3)	36
Poland	(6.4)	(2)	(6.0)	(1)	−6
Persia	(2.9)	(1)	(10.1)	(2)	249

Note: a. Because of rounding the total may not equal the sum of the column.
Source: H. Williamson, R. Andreano, A. Daum and G. Klose, *The American Petroleum Industry, The Age of Energy, 1899-1959* (Northwestern University Press, Evanston, Ill., 1963), p.262.

skyrocketed during those years, rising from 26 million barrels in 1914
to 87 million in 1919. By the latter year Mexico replaced Russia as
the world's second largest producer.[43] In that year Mexico produced
about one-sixth of the world's oil. Its rise in production was greater
than any other country with the exception of Persia. In that country,
production expanded sharply to meet the war-time needs of the British
Navy. Nevertheless it was still relatively small, only 10 million barrels.

During this period Mexico's exports to the US rose from 18 million
to 53 million barrels or by 200 per cent. By 1919 these exports
equalled 14 per cent of US production.[44] By any measure, Mexican
oil was very important both to the US oil industry and the US
economy.

It is therefore not surprising that the US Government reacted with
extreme hostility towards Mexico when Carranza, the man whom it
put in power, moved against the oil industry. Carranza proposed a new
Constitution which would have nationalised sub-surface mineral
resources. This struck a direct blow to those oil companies who had
received concessions but had not commenced drilling over the entire area
of their concessions. The great majority of companies were in this
category. The Mexican Government generally refused to issue drilling
permits to these companies unless they acknowledged the nation's
ownership of the sub-surface resources.

The US Government protested against these actions, claiming that
they were retroactive. Mexico, for its part, denied the retroactivity
charge, contending that the nation was merely reclaiming the resources
which were illegally ceded to the companies by Diaz. The Mexican
Government also pointed out that, as a sovereign nation, it had the
right to dispose of its resources in any way it saw fit.

In reply the United States set forth its doctrine of 'minimum duty'
in a note sent to the Mexican Government in 1918. Henry Fraser,
Chief Counsel to the US Senate Special Committee Investigating
Petroleum Resources, later summarised the US position as follows:

The principal point made by the United States . . . was that every
nation has certain minimum duties to perform with regard to the
treatment of foreigners, irrespective of its duties to its own
citizens, and that in default of such performance it is the right of
the foreign government to enter protest. Since one of the duties
imposed upon a government is to refrain from measures resulting
in confiscation of the vested property rights of foreigners, the
United States is not estopped, the note said, from protesting on

behalf of its citizens merely because the Mexican Government may choose to confiscate vested property rights of its own citizens. The note concluded with the hope that the Mexican courts would protect the legitimately acquired rights of American citizens and thus happily end the controversy; but if this hope should be disappointed, 'the Government of the United States must reserve to itself the consideration of the questions of interesting itself on behalf of American citizens concerned in this important and serious matter'.[45]

There is no question that this note threatened armed intervention unless the Mexican Government adopted the doctrine of 'minimum duty'. The Mexican Government under President Calles finally acceded to the US demands in 1928.[46] This US victory was relatively short-lived as Mexico finally nationalised the oil properties in 1938.[47] The implications of this nationalisation will be discussed later. What is important at this point are the implications of 'minimum duty'. This doctrine in effect said that foreigners − or, more precisely, foreign investors − are a special, privileged class, not subject to the laws of the nation in which they do business. Any nation could enact any legislation it wished regulating the business activities of its citizens so long as these laws did not apply to foreigners. This doctrine served the convenience of the United States in its dealings with Mexico. It did not however serve the interests of the United States in its dealings with England and Holland in the post World War I period. The US used the doctrine of the 'open door' against them.

Open Door

That these two doctrines conflicted with each other never seemed to bother US policy-makers. They used whatever doctrine served their purpose at the moment. The doctrine of 'minimum duty', with its provision of special privilege, was suitable for a country in which the US had already acquired a large vested interest. The doctrine of 'open door', with its implications of non-discrimination, was, on the other hand, suitable for an area from which US interests were being excluded. In any event both doctrines served the goal of protecting and expanding US oil interests abroad.

Iraq. The US first used the 'open door' doctrine as a means of forcing the British to give US oil companies the right to participate in the exploitation of Iraq's oil resources. The British, after long and often acrimonious discussions, finally agreed to let the American companies in.

The dispute centred on the post-war reorganisation of the TPC (Turkish Petroleum Company). This company, formed in 1912 when Turkey held suzerainty over Iraq, represented a combination of British and German interests put together by C.S. Gulbenkian, a Turkish Armenian, to exploit the petroleum resources of the provinces of Mosul and Bagdad. The share-holders of the company were: the National Bank of Turkey, which was owned entirely by British interests; Anglo-Saxon Petroleum Company, a subsidiary of Royal Dutch Shell; and the Deutsche Bank. The shares were divided as follows: 25 per cent each for Anglo-Saxon and the Deutsche Bank and 50 per cent for the National Bank of Turkey, of which 15 per cent was held for Gulbenkian.[48] Early in 1914 the National Bank's holdings were transferred to Anglo-Persian with the stipulation that 5 per cent was to be set aside for Gulbenkian. That 5 per cent was to be shared equally by Anglo-Persian and Anglo-Saxon.[49] Thus Anglo-Persian wound up controlling 47.5 per cent of the shares; the Deutsche Bank, 25 per cent; Anglo-Saxon, 22.5 per cent; and Gulbenkian, 5 per cent.

The outbreak of World War I brought all operations of the TPC to a halt. The British Alien Property Custodian took over the interests of the Deutsche Bank. After the war the British and French signed an agreement in San Remo transferring the Deutsche Bank's shares to the French Government. The agreement also permitted the government of Mesopotamia (now known as Iraq), over which the British were given a mandate, to receive a 20 per cent share in the Company.[50] No US company was given any share in TPC.

Around the same time seven American companies[51] wanted to send geological expeditions into the country. The British, as the mandatory power, refused to grant them permission to enter. At this point the US State Department entered the picture. A note delivered by the US Ambassador to Lord Curzon, the British Foreign Minister, complained that Britain 'had given advantage to British oil interests which were not accorded to American Companies and further that Great Britain had been preparing quietly for exclusive control of the oil resources in this region'.[52]

The note demanded that 'as a general principle any Alien territory which should be acquired pursuant to the Treaties of Peace with the Central Powers must be held and governed in such a way as to assure equal treatment in law and in fact to the commerce of all nations'.[53] What the US was in effect demanding was that it, as an

Allied power, be treated at least on an equal basis as the French. The problem was that the US, unlike Britain and France, never declared war on Turkey and never was a signatory to the peace treaties concluded with Turkey.[54] Furthermore, the mandate system by which Mesopotamia was governed, was under the control of the League of Nations. The United States refused to join the League. The legal bases for its claims in this region were therefore rather tenuous.

In answer to these objections, Secretary of State Colby replied:

> While it is true that the United States of America was not at war with Turkey, yet it was at war with the principal allies of that country and contributed to the defeat of those allies and, therefore, to the defeat of the Turkish Government . . . The United States, as a participant in that conflict and as a contributor to its successful issue, cannot consider any of the associated powers, the smallest not less than itself, debarred from the discussion of any of its consequences, or from participation in the rights and privileges secured under the mandates provided for in the treaties of peace . . .[55]

In another note the US reiterated the 'open door' doctrine, which, in this case, meant equality of treatment for American interests. The US demanded:

> That no exclusive economic concessions covering the whole of any Mandated region or sufficiently large to be virtually exclusive shall be granted and that no monopolistic concessions relating to any commodity or to any economic privilege subsidiary and essential to the production, development, or exploitation of such commodity shall be granted.[56]

It is interesting to note, in this respect, that in its communications with Mexico the US never expressed any concern about the monopolistic practices of its companies in that country. The US note further stipulated that:

> in general regulations or legislation regarding the granting of concessions relating to exploring or exploiting economic resources or regarding other privileges in connection with these shall not have the effect of placing American citizens or companies or those of other nations or controlled by American citizens or nationals of

other countries at a disadvantage compared with the nationals or companies of the Mandate nation or companies controlled by nationals of the Mandate nation or others.[57]

There was no mention here of the 'minimum duty' to the original concession holders, a point which was the main concern of the United States in its negotiations with Mexico. In any event, the US finally reached an agreement with England, which allowed it, Great Britain, Holland and France to have access to Iraq's petroleum resources to the exclusion of everyone else.

The US was able to gain entry for its companies because it controlled most of the world's oil supply. It was in a position to withhold these supplies from recalcitrant nations. The US Federal Trade Commission noted:

> The American oil industry produced in 1921 about 65 per cent of the world's oil supplies, and purchased about 17 per cent of the remainder (mostly from Mexico). After satisfying all American needs, the industry exported about 58 per cent of total foreign requirements. Anglo-American Oil Co., a former subsidiary of Standard Oil Co. (New Jersey), controlled over 50 per cent of the total business of the United Kingdom. Thus, the American oil companies were in a strong bargaining position. It is not known if they actually used their bargaining strength to the extent of threatening to withdraw from the export market, but the possibility was discussed by officials of Standard Oil Co. (New Jersey).[58]

It should be noted that no one need explicitly mention such a threat. As long as those on both sides of the bargaining table know that the potential of withdrawal exists, they will adjust their strategies accordingly. Thus, the control of oil carried with it not merely the ability to wage war, which so concerned Churchill and Requa, but also the potential to impose one's will on other nations. This potential was explicitly recognised, as we shall see later, by the US in the post World War II period.

The final agreement, reached in 1928, granted to the NEDC (Near East Development Company), a corporation established to handle the American interests, a 23.75 per cent share of TPC. This share was equal to that held by each of the other major partners, Anglo-Persian, Royal Dutch Shell and Compagnie Française des Pétroles. The remaining 5 per cent was allocated to Gulbenkian. Standard Oil (New

Jersey) and Standard Oil (New York) each received a 25 per cent
interest in NEDC. Three other companies — Gulf, Atlantic and
Pan-American Petroleum and Transport, a subsidiary of Standard Oil
(Indiana) — each obtained 16⅔ per cent.[59] Subsequently each of these
three sold their shares to the two major partners so that by 1934, these
two firms held an equal interest in NEDC.[60]

These results show how drastically the attitude of the US
Government had changed towards the Rockefeller interests.
Rockefeller and his companies were transformed from pariahs into
key instruments of US foreign policy. With the full blessing of the
State Department the negotiations between the seven US oil
companies and the TPC were conducted by W.C. Teagle and W.D.
Asche, president and vice-president of Standard Oil (New Jersey).[61]
Furthermore, by backing the entry of supposedly competing companies
into the TPC, the State Department put these companies in
a position where they would have to enter into production-sharing
arrangements. Such arrangements violated the essence of the US
anti-trust laws, the very laws which the US Government had used
against Standard Oil two decades earlier. In addition, all the partners
in TPC had to abide by the 'Red Line Agreement' which forbade each
one of them to bid individually on concessions in the general area
controlled by the old Turkish Empire.[62] All such bids were to be
handled through TPC. This stipulation further cartelised the industry
and reinforced the tendency for the TPC owners to work together
rather than to compete with each other in the market place. Finally,
the TPC agreement had the effect of enhancing the power of the
Rockefeller-controlled companies by giving them access to prolific
reserves of crude. As mentioned in chapter 3, the Rockefeller
companies were traditionally short of crude because of Rockefeller's
aversion to the purchase of reserves. The US Government, through
its actions on the Mesopotamia issue, helped the Rockefeller
companies to overcome this disadvantage. It also helped the
Rockefeller interests to obtain reserves in the Netherlands East Indies.

Netherlands East Indies. Before World War I practically all the oil in
the Dutch East Indies was controlled by Royal Dutch Shell either
directly or through its subsidiaries. The Dutch Government ostensibly
did not discriminate against non-nationals. Anyone was allowed to
apply for an exploration permit and would receive a concession if oil
were found. In practice the authorities were reluctant to grant exploration
permits to foreigners, thereby giving a virtual monopoly to Shell.

Despite these obstacles, several foreign companies set up subsidiaries in the Indies with the hope of eventually receiving concessions. Among them was a subsidiary of Standard Oil (New Jersey), the Nederland Koloniale Petroleum Maatschappij, which was organised in 1910.[63] During the war the Dutch Government amended its mining laws limiting prospecting licences or concessions to:

(a) Netherlands nationals; (b) inhabitants of the Netherlands or of the Netherlands East Indies; (c) companies domiciled in the Netherlands or in the Netherlands East Indies, of which, in the case of limited companies, the majority of the managers or directors were Netherlands nationals or inhabitants of the Netherlands East Indies.[64]

While the US was somewhat perturbed by this provision, it did not regard it as a serious obstacle to US companies. These companies could, and probably would, require that the managers and directors of their subsidiaries become residents of the Indies. This provision therefore did not seriously conflict with the concept of the 'open door'. Far more disturbing was another provision, Article 28, which severed the link between the discovery of oil and the right to a concession. Under the old mining law, it should be recalled, anyone who made a discovery would automatically receive a concession to exploit it. The new law, on the other hand, stated that 'only the Government, or companies with which the Government entered into contracts . . . would have the right to extract oil'.[65] The US Government correctly interpreted this article as a device to exclude non-Dutch companies.

The issue came to a head after the discovery of the Djambi fields in Sumatra. The US Consul at Batavia described these fields as 'the most valuable mineral oil fields in the whole colony'.[66] In 1920 the Dutch Government, eager to keep this find under Dutch control, introduced a Bill into the States-General granting the entire area to a new company, equally owned by the government and Bataafsche Petroleum Maatschappij, a subsidiary of Shell. This Bill thus effectively excluded American interests from the area.

During the debate over the Bill the US expressed its concern to the Dutch Government. In the initial discussions the US Minister to the Netherlands did not discuss the Djambi Bill but the revisions to the mining law. While admitting that these revisions conformed to the 'open door' concept, he objected to them on the grounds of 'reciprocity'.

According to Fraser, the State Department complained that 'the Netherlands had not accorded the same degree of freedom to American nationals as was accorded to Dutch nationals in the United States' and therefore 'had not granted reciprocal privileges to American capital'.[67] The Department also noted that, while the Dutch laws apparently did not discriminate, they were administered in such a way as to exclude American firms. Reciprocity, the Department contended, 'must be not only in the letter of the law but in its administration and operation'.[68]

In the meantime two American companies, Sinclair and Koloniale, Standard Oil's (New Jersey) subsidiary, applied for concessions in Djambi. Their applications were turned down. The US Government protested vigorously. In a note to the Dutch Government the American Minister stated the essence of US policy:

> the United States has for years carried the burden of supplying a large part of the petroleum consumed by other countries, that Dutch capital has had free access to American oil deposits and that the petroleum resources of no other country have been so heavily drawn upon to meet foreign needs as the petroleum resources of the United States. I have pointed out in the future ample supplies of petroleum have become indispensable to the life and prosperity of my country as a whole, because of the fact that the United States is an industrial nation in which distance renders transportation difficult and agriculture depends largely on labor-saving devices using petroleum products.[69]

This note was a much more explicit formulation of the policy of oil imperialism than that of Requa. The emphasis had however shifted somewhat. While Requa was concerned about the military uses of oil, the American Minister to the Netherlands stressed the importance of oil for the economic development of the US. He, like Requa, was not content to let the US consumers of oil buy that commodity in the market. The reserves of oil, wherever and in whatever country they existed, had to be controlled by US companies. He emphasised this particular point by informing the Dutch that 'American capital stands ready to assist in the development of the Djambi fields and other oil deposits in the Netherlands Indies'.[70]

The Dutch protested that if the principle of 'reciprocity' were universally applied they would have to divide up the Djambi fields among many nationalities. But these protests fell on deaf ears. The

US invoked the 'non-reciprocity' provisions of the General Lease Act against Shell by refusing it a permit to explore on Federal lands in Utah. Finally, in 1927, the Dutch Government introduced a Bill to the States-General authorising the issuance of four tracts of land for exploration and production to Koloniale. The government also promised to continue this policy on the condition that the US declare Holland a 'reciprocating country'.[71] Such a declaration would again enable Shell, or any other Dutch company, to explore on Federal lands.

Several days later the US Government informed the Netherlands that it had accorded Holland recognition 'as a reciprocating State within the meaning of the provisions of the Mineral Leasing Act'.[72] Thus, by using the hallowed principle of 'reciprocity', the US Government forced the Dutch Government to open its door to Rockefeller's Standard Oil (New Jersey).

As an aside, it might be interesting to note how seriously the US viewed 'reciprocity'. At the same time as the State Department was negotiating with the Dutch, the US Navy was formulating policy for the use of its oil reserves. The Navy wanted to insert a clause, similar to that in the Mineral Leasing Act, restricting concessions to nationals from 'reciprocating' countries. The State Department cautioned against the insertion of such a clause, arguing that 'it could hardly be of any great practical importance' since the 'naval petroleum reserves are less than 1½ per cent of our total petroleum reserves, and the larger part of the naval reserves has already been leased under long term contracts', affording only a very limited opportunity 'for foreign-controlled companies to participate in the further development of these reserves'.[73] This was a rather revealing statement from an agency which, at the same time, was demanding that the Dutch observe 'reciprocity' in 'administration and operation' as well as in 'the letter of the law'.

The State Department had more basic reasons to oppose the insertion of a 'reciprocity' clause. First of all, it feared that such a clause 'might tend to increase agitation in this country against the operation of foreign petroleum interests in the United States', thus tending 'to make more difficult the support by this Department of American petroleum interests abroad'. In addition the Department was afraid that a 'reciprocity' clause would induce other nations also to establish naval reserves and use this clause to exclude US interests. The Department was particularly worried about the effects on Brazil, which was 'considering a proposal which would declare all petroleum deposits as well as deposits of many of the most important minerals to be reserved for national security and defense and excluding foreign

interests from opportunity to participate in their development'.[74]
'Reciprocity', like 'minimum duty' and 'open door', hence turned
out to be a disposable weapon in the US diplomatic armoury. They
were to be used when needed and then discarded when they stood
in the way of attaining the objective of ensuring American control
over the world's oil resources.

The US Government promoted the expansion of the American
oil industry in many parts of the world. The interwar period
witnessed a spectacular growth in the overseas operations of US oil
companies, especially in Latin America and in the Middle East.
Sometimes this expansion was aided by direct State Department
pressure. At other times such pressure was not necessary because the
leaders of the host country welcomed the investment inflow. But in
all countries the leaders knew that behind the oil companies lay
the full power of the United States Government.

Venezuela. In Latin America, Venezuela was the most important
country for the US oil industry. Large-scale investments started
flowing there after the drilling by Shell of the Barrosa No. 2 well in
the Lake Maracaibo area in 1922. The well came-in so strongly that
it 'blew-out'. This 'blow-out', which spewed forth 850,000 barrels
of oil, lasted nine days. The news of this find electrified the oil world.
Within days oil men from all over the world were scrambling to obtain
concessions in the Lake Maracaibo area.[75]

For most the trip to Venezuela was in vain. They could not
obtain concessions because Shell held a virtual monopoly of the
on-shore fields. The only concessions available were those on the floor
of the lake. Few companies felt that these concessions were worth
bidding for.[76] Two companies, Gulf and Standard of Indiana,
however, thought that they could drill on the lake floor. Gulf obtained
concessions in the area along the shore and Indiana in the deeper areas.
Since the pool discovered by Shell extended under the lake, these
companies were also able to drill highly prolific wells.[77]

These two concessions were the first major American inroads into
Venezuelan oil. Up until 1923, when the Americans entered in force,[78]
the main companies in Venezuela were Shell and British Controlled
Oilfield, a company controlled by the British Government.[79] This
control had to be kept secret because Venezuelan law forbade the
granting of concessions to government-owned companies.[80] Shell
obtained its foothold in 1911 through the acquisition of General
Asphalt, an American company which had obtained a concession to

the area lying east of Lake Maracaibo.[81] British Controlled Oilfield received its concession in 1917.[82]

The entrance of Shell into Venezuela initially did not disturb the Rockefeller oil interests. Venezuela was regarded as a source of asphalt rather than of crude. Furthermore, Standard's agents felt that the country was too unhealthy a place.[83] This attitude changed however when it became apparent that the country had a large oil potential.

One of the problems which Standard Oil (New Jersey) and the other entrants had to face was how to break the British monopoly on concessions. The British obtained their holdings under the regime of President Juan Vincente Gomez, who seized power in 1908 with the help of the US Navy.[84] Gomez had the habit of granting concessions of mineral resources, which by Venezuelan law belonged to the nation, to his political favourites. They, in turn, resold their concessions to the oil companies. Apparently, Gomez himself also exacted a charge for these concessions. When he died in 1935, he left an estate valued at $200 million.[85] In any event, Shell, ever anxious to find new reserves, took advantage of this opportunity to obtain concessions at a very low price. That Shell benefited enormously from this system is evident from the praise Sir Henri Deterding, the head of Shell, lavished on Gomez shortly before the latter's death.

> I satisfied myself . . . that the government under General Gomez appeared sound and constructive and likely to be fair to foreign vested interests. And now that I know Venezuela better, I can testify that in his 26 years of virtual dictatorship, General Gomez has consistently insisted on fair play to foreign capital.[86]

In order to break the British monopoly, the US State Department induced the Venezuelan Government to force Shell to surrender the concessions it had not explored.[87] These lands were, in turn, sold to Gomez's supporters, who resold them to the US oil companies. The State Department also informed the Venezuelan Government that a British company, presumably British Controlled Oilfield, was controlled by the British Government and, therefore, not entitled to concessions.[88] The US Minister arranged, on State Department orders, a private meeting between Gomez and representatives of Sinclair Oil.[89] Most important of all was the role played by the US Government in revising Venezuela's basic petroleum law. In 1921, the US Minister co-operated with Standard Oil (New Jersey), Sun Oil and other companies in drawing up proposals to change the law in order to 'make it possible for foreign oil companies to extend their oil operations'.[90] Their proposals were incorporated in the Petroleum Law of 1922, which the US oil companies regarded as the best in Latin America.[91]

Following the enactment of this Law, a number of important companies set up operations in Venezuela. They are listed in Table 4.3.

Table 4.3: Major American Companies Entering Venezuela after Passage of 1922 Petroleum Law

Company	Controlling interest	Year of entry
Orinoco Oil	Pure Oil	1923
Venezuelan Gulf	Gulf	1923
Lago Petroleum	Doheny[a]	1923
Richmond Petroleum	Standard Oil of California	1925
Venezuela Atlantic Refining	Atlantic Refining	1926
California Petroleum	Texas Company	1927

Note: a. Standard of Indiana purchased Lago from Doheny in 1925.
Sources: L. Vallenilla, *Oil: The Making of a New Economic Order, Venezuelan Oil and OPEC* (McGraw-Hill, New York, 1975), p.37; Luis Alcala Sucre, 'The Impact of Multinational Oil Companies in Venezuela: The Oil Companies in General and Gulf Oil in Particular' in Jon P. Gunneman (ed.), *The Nation-State and Transnational Corporations in Conflict with Special Reference to Latin America* (Praeger, New York, Washington and London, 1975), p.184.

Standard Oil (New Jersey) and Sun were the only important US companies with holdings in Venezuela before the enactment of the Law. Standard entered the country in 1921 through a subsidiary, Standard Oil Company of Venezuela.[92] According to Harvey O'Connor, it was able to pick up its early concessions only by learning the 'rules of the game' from Sun. The company approached Gomez with a bid on five concessions. Gomez's son-in-law also bid on those concessions, obtained them and promptly resold them to Sun. From that day on, Standard dealt directly with Gomez's son-in-law.[93]

The turmoil in Mexico also played a role in the movement of many US firms to Venezuela. The persistent agitation for the nationalisation of sub-surface mineral rights caused many firms to divert investment from Mexico to Venezuela. Furthermore, the companies in Mexico pumped the fields as quickly as they could, thereby rapidly depleting the nation's oil. Production, after peaking in 1921, started declining. The discovery of prolific oil areas in Venezuela offered a welcome substitute to Mexico. Thus the rise of oil in Venezuela was accompanied by the decline of oil in Mexico.

The Mexican events were an important factor causing Doheny to move to Venezuela. His Venezuelan subsidiary, Lago Petroleum, obtained prized concessions on the floor of Lake Maracaibo. In 1925 he sold his holdings to Standard of Indiana,[94] making Indiana one of the major companies in the country. By 1929, a year after Venezuela became the second largest producer of oil in the world,[95] three companies – Shell, Gulf and Indiana – dominated the scene. Shell had 45 per cent of production, the other two companies had 27 per cent each. Venezuelan oil provided one-half of the production of Shell and Gulf and more than one-half of Indiana's.[96]

The developments in Mexico also affected the location of refineries. Despite Gomez's generosity to the oil companies, the companies feared that the same nationalist sentiments which swept Mexico could sweep Venezuela. As a precaution, Shell decided to build a giant refinery on the Dutch-owned Caribbean island of Curaçao. By doing so, it initiated a policy that was to be followed by all the other international oil companies, that of locating refineries outside the political jurisdiction of the host countries. Since crude oil is worthless unless it is refined, the host countries cannot sell their crude if they have no refineries. Thus the locational policy of the oil companies greatly enhanced their bargaining power *vis-à-vis* the host countries. If the latter either increased their royalty demands or nationalised the fields, the companies would merely refuse to refine the crude.

Standard of Indiana followed Shell by constructing a major refinery on another Dutch Caribbean island, Aruba. The Dutch apparently were reluctant to grant permission for this refinery, which would compete with Shell. They were however then involved in the dispute with the US over Djambi and did not want to antagonise the US further by denying Standard of Indiana permission for the refinery.[97] This refinery, whose main market was the US, began operations in 1929.[98] The main market of the Shell refinery was Europe.[99] These two refineries processed about one-third of Venezuela's crude.[100]

Indiana's refinery on Aruba opened at a particularly unpropitious time, the eve of the Great Depression. The reduction in demand brought downward pressure on prices. Accompanying this decline in demand was a significant increase in supply following major discoveries in Oklahoma and in East Texas. This combination of decreased demand and increased supply severely depressed prices in the States. Added to this was the downward pressure of increased imports from Venezuela. Soon a call arose for the restriction of imports. The US Government in 1933, as part of the NIRA (National Industrial Recovery Act) programme, imposed controls on imports, limiting them to 'the daily average of petroleum and petroleum products during the last six months of 1932'.[101] These controls struck a virtual death blow to Standard of Indiana's foreign operations. Since Indiana had no marketing facilities outside the United States, it could not divert its foreign production to other markets. In anticipation of these controls, it sold its Venezuelan properties, including the refinery on Aruba, to Standard Oil (New Jersey) in 1932.[102] Jersey then exported refined products from Aruba to Europe, thereby complying with the US import programme. At the same time it reduced its refined product exports from the US.[103]

Following this acquisition, Jersey, which in 1929 accounted for less than one per cent of Venezuela's production, became by 1935 the largest company in Venezuela, producing more than one-half of total output.[104] By the end of the 1930s, Jersey, Shell and Gulf controlled the bulk of Venezuela's production. Venezuela was the most important oil producer outside of the United States. Its importance was explicitly recognised by Nelson Rockefeller, who in 1935 deliberately put himself on the board of Jersey's subsidiary, Creole.[105] He regarded it as the plum of the Rockefeller Empire since it was the purchase of the Venezuelan properties that made Jersey the largest oil company in the world.[106] Despite its early lack of success in Venezuela, Jersey eventually became the kingpin of the Venezuelan economy. In this it was aided, at least indirectly, by US Government policy.

Beside Venezuela, US firms entered a number of Latin American countries, among them Peru, Bolivia and Colombia. Petroleum investment in all of Latin America significantly expanded during the interwar period. This expansion however was not a continuous one. In two countries, Bolivia and Mexico, investment fell because of expropriations. Bolivia expropriated the properties of Standard Oil (New Jersey) in 1937.[107] Mexico nationalised its oil in 1938.[108]

The Mexican nationalisation occurred because the oil companies, by refusing to obey an order of the Mexican Supreme Court, forced the government to act against them. The government at that time had no policy for nationalisation. It was content to let the companies operate in peace. It was however forced to intervene in a strike against the companies. It appointed a number of bodies to investigate the workers' grievances. These bodies upheld the workers' demands, especially those for wage increases, on the grounds that the companies

were extremely profitable and could well afford to pay them. The
companies appealed against these findings to the Supreme Court. The
Court sustained the findings. Whereupon the companies refused to
obey the Court because they said they could not afford to do so.
Following this refusal the government expropriated the properties.[109]
The companies affected were: Shell, Standard Oil (New Jersey), Sinclair
and Standard Oil of California.[110]

The companies retaliated by imposing a boycott on Mexico. They
threatened all companies buying Mexican oil with law suits, including
those companies selling machinery to the Mexican oil industry. They
refused to sell Mexico any materials and the Mexican Government
could not obtain replacement parts.[111] They also asked their home
governments to exert diplomatic pressure on Mexico. While both the
United States and Britain protested vigorously, they did not go
beyond making these formal protests. In particular, the United States
did not use military force against Mexico. There were a number of
reasons for this moderation in US policy. First, the US Ambassador
to Mexico, Josephus Daniels, was sympathetic towards the Mexican
Government. He blamed the expropriation entirely on the ineptness
of the companies. In a letter to Secretary of State Cordell Hull, who
wanted to take a more vigorous approach towards Mexico, Daniels
wrote: 'I wonder at the ineptness of the representatives of the oil
companies . . . Some are so dumb that if they had to start a business
on their own, it would be foredoomed to failure. Initiative and tact
are not in their vocabulary . . .'[112] Daniels' arguments were convincing
enough to dissuade Hull from intervening militarily.

Second, Mexico was no longer important in the world oil picture.
Its production had fallen from its peak of 193 million barrels in 1921
to 47 million barrels in 1937. Of this latter amount only 54 per cent
was exported while in the heyday of Mexican production over
80 per cent was exported.[113] Thus the nationalisation did not have a
major impact on the world oil industry.

Third, the US Government under President Franklin Roosevelt was
anxious to improve its relations with Latin America. Officially the
'Good Neighbor Policy' displaced 'Dollar Diplomacy'. Intervention in
Mexico would have destroyed this new policy. Furthermore, it was
feared that it might spur demands for nationalisation in Venezuela,
which was far more important both to the oil industry and the US
Government. Intervention in Mexico was not worth taking such a risk.

When it became apparent to the oil companies that the US
Government was not going to intervene, they began to change their

tune. Initially they refused to consider any offer by Mexico for compensation, demanding instead the return of their properties. Finally, in 1940 Sinclair accepted an offer of $8.5 million. Others followed suit. All told, the Mexican Government paid out $165 million in principal and interest.[114]

The Mexican episode illustrates the importance to the oil companies of backing by their home governments. With such backing they can treat the host governments as colonial lackeys, whom they can order around at will. Without such backing they have to accede to the demands of the host governments.

Middle East. While in terms of production and investment Latin America was the most important area of expansion for US oil companies in the interwar period, the Middle East also attracted their attention. This area was destined to replace Latin America as the key oil area in the post World War II era. As has already been discussed, the first entry of US firms into the Middle East was in the TPC, which changed its name in 1929 to IPC (Iraq Petroleum Company).[115]

Around the same time as the US was negotiating to allow American companies in Iraq, the Gulf Oil Company obtained an option to purchase concessions in Bahrein and Kuwait from Eastern and General Syndicate, a British company. This company, headed by Major Frank Holmes, a New Zealander, obtained these concessions in 1925. When Gulf was ready to exercise its options, it had already joined IPC and was bound by its 'Red Line' agreement, which prohibited IPC owners from bidding on their own for concessions in most of the territory of the old Ottoman Empire. Bahrein was within the 'Red Line' but Kuwait was on the outside.

Gulf had to ask the IPC for permission to exercise its option in Bahrein. The IPC, fearing the entry of another American firm into the area, refused. Gulf then sold its option to Standard Oil of California, which was not a participant in IPC and therefore not bound by the 'Red Line' agreement.[116] The British, who exercised control over Bahrein, tried to prevent Standard of California from obtaining the concession. The US State Department intervened and the concession was approved in 1930.[117]

Although, because Kuwait was outside of the 'Red Line', Gulf did not need approval from IPC to exercise its option there, it was blocked by the British Government from doing so.[118] The British insisted that the concession be given to a British company. The State

Department protested, demanding that US companies be given equal treatment. At about the same time Anglo-Persian, a partner in IPC, put in a bid for a concession. After much haggling, the two companies decided to submit one bid. In 1933 they formed the Kuwait Oil Company, in which each held an equal share, to operate their Kuwait concession.[119]

Saudi Arabia was the next important area of American penetration in the Middle East. As in the rest of the Middle East, the Americans were not the first to enter. Major Holmes' Eastern and General Syndicate obtained a concession from King Ibn Saud in 1923.[120] Though the US Government had no direct interest in that area and had no diplomatic or consular officers in the country, it nevertheless instructed its diplomats elsewhere in the Middle East to obtain the details of the concession. The American Vice-Consul in Bushire, Persia reported back the details which, he claimed, were furnished him by 'a British subject, the agent of Standard Oil'. He warned that this 'source in particular, should be kept confidential'.[121] The Vice-Consul also noted that 'the concession is reported to be with the consent of the Colonial Office'.[122] He then commented:

> It is generally agreed that there is no real competition between the Eastern & General Syndicate, and the Anglo-Persian Oil Company. A. T. Wilson [General Manager of Anglo-Persian] tried to get the same concession for the latter company, and also tried to get a concession in Koweit [*sic*] His failure seems not to worry anyone, and may have been more or less deliberate.[123]

Sir Arnold Wilson apparently believed that there was no oil in Saudi Arabia. So also did King Ibn Saud. He accepted advances from Eastern & General never believing that oil would be found.[124]

Saud's judgement was correct as far as Eastern & General was concerned. Holmes's company found no oil and let the concession lapse. The increasing demands of Ibn Saud left Holmes no other choice. A 1926 dispatch from the American Consulate in Baghdad indicates the nature of some of these demands:

> I learn that Major Frank Holmes considers that his oil concession . . . has lapsed. It is said that, in order to get this concession in the first place, Major Holmes was obliged to pay Ibn Saud a considerable sum of money in cash and also to give him an expensive motor car . . . now Ibn Saud is demanding more and further presents from

Major Holmes if the concession is to remain in force. I am informed that apparently the money and presents are not forthcoming and it was when speaking about the new demands of Ibn Saud that Major Holmes remarked that his concession had lapsed.[125]

After several years Standard Oil of California started negotiating with Ibn Saud for the lapsed concession. Though the State Department did not directly participate in the negotiations, it kept close tabs on developments. State Department participation was not necessary because King Saud, unlike most of the other Middle East potentates, was not under British control. The British backed his rivals, the Hashemites, in the battle with Saud's Wahabis in the post World War I struggle for control of the country. Saud was therefore very receptive to American investments as a counterweight to British influence in the Middle East.

An important negotiator with King Ibn Saud was Kenneth Twitchell, an independent American geologist, who acted on behalf of Standard Oil of California. He also kept the State Department informed about the negotiations. A State Department memorandum of 1932 summarised Mr Twitchell's impressions as follows:

> Mr. Twitchell said that the present was an excellent time to enter the scene in Saudi Arabia. King Ibn Saud, he asserted, was disposed to be very friendly to the United States and was hopeful that Americans, who had no political ax to grind, would become interested in the exploitation of the natural resources of his Kingdom. The pilgrimage business, from which Saudi Arabia derives most of its revenue, has been seriously hit by the depression, and the revenue that might be expected from oil or other mineral development would be especially welcome at this time.[126]

As an indication of the struggle for control over Middle East oil resources, Twitchell also discussed the attitudes of the foreigners in Saudi Arabia towards the United States. The State Department summary states that Twitchell described the Dutch as 'the most friendly' while the British 'would be sure to look with suspicion and disfavor on the entrance of American interests'.[127]

Recognising the British antagonism to American entry into the area, the State Department tried to adopt a position between giving in to British desires to exclude American influence, on the one hand, and from ousting Britain as the dominant power in the region, on the other.

This ambivalent position was evident in a reply to a telegram sent to the Department in 1932 by Francis B. Loomis, representative of Standard Oil of California. After informing the Department that his company was about to sign a contract with King Ibn Saud, he asked, 'What protection could we expect in case of disorders in Arabia or in case the present government is overthrown and its successor were disposed to break our contract without just cause?'[128]

The Department's reply was ambiguous. It stated:

> The Department is unable to indicate in advance the nature of
> the protection, if any, which it could accord in the event of
> the contingencies to which you refer, but you will doubtless
> appreciate the possible difficulty of assuring effective protection
> for American influence.[129]

The 'difficulty' presumably would arise from the dispatch of a military force to an area regarded as in the British sphere of influence. The Middle East, after all, was not Latin America.

Despite the lack of a firm guarantee of protection from the US Government, the company went ahead with its negotiations. In May 1933 it signed an agreement with the government of Saudi Arabia. This agreement, in addition to stipulating the area and time period of the concession, the terms and methods of payment, and many technical details, provided that 'the enterprise . . . shall be directed and supervised by Americans'.[130] In other words, this Saudi oil was to be considered as American oil.

Apparently the terms of the agreement were highly favourable to the company. For instance, it exempted the company from paying duties on its imports and gave the company the right to use, free of charge, the government's water and other natural resources.[131] Though the agreement called for publication 'in Saudi Arabia in the usual manner',[132] it was not published in its entirety. In a letter to the State Department Twitchell noted that the 'Government is not publishing details of the loans and payments, as it wishes to avoid the criticism of those who might say it should have obtained terms equal to the Iraq and Persian governments'.[133]

In any event the company was not able to enjoy any profits from its investment until 1938, when it discovered oil in commercial quantities.[134] In the meantime it brought in a partner, the Texas Company, in 1936. They created ARAMCO (Arabian-American Oil Company), a company in which each held a 50 per cent share.

Standard of California brought in Texas because it needed marketing facilities for its Bahrein production. It discovered oil there in 1932. It was reluctant to sell its oil in the States, where it had marketing facilities, because its Bahrein oil would compete with its California oil. Since it had no marketing facilities outside the States, it tried to make an arrangement with the Big Three — Shell, Anglo-Persian and Standard Oil (New Jersey) — who controlled most of the world's marketing facilities. But no agreement could be reached. It then turned to Texas, which had extensive marketing facilities in Europe, China, Australia, Africa and other areas of the Far East. Texas was supplying these areas from the United States but found it more profitable to supply them from the Middle East. Therefore it had an incentive to enter into arrangements with Standard of California.[135]

With the entry of American companies into Saudi Arabia, the US had an interest in every important Middle East producing area with the exception of Iran. That country was completely controlled by the British, who were successful in excluding everyone else during the interwar period. For many years that country was the only producer of oil in the Middle East, giving the British a virtual monopoly of Middle East production. This monopoly lasted throughout the 1920s. During the 1930s production commenced in Iraq, Bahrein and Saudi Arabia. As a result the British share of Middle East production fell and the non-British share, especially the American, rose.

A similar pattern prevailed in the Far East. In this area the dominance of the British-Dutch firms eroded as the American share rose. Only in the Western Hemisphere did the American share fall. This was due mainly to the nationalisations in Bolivia and Mexico and the insistence by some Latin American countries that their own nationals be permitted to participate in oil operations. The fall in the US share was moreover not accompanied by a rise in the British share but rather by a rise in the share held by Latin American nationals. To a certain extent this meant a rise in the US share since some, but by no means all, Latin American shareholders were subject to US influence and control. These control patterns are summarised in Table 4.4.

These changes in control patterns followed the changes in investment patterns. As has already been shown in Table 4.1, practically all investment in production in 1919 was in the Western Hemisphere, with more than three-fifths concentrated in Mexico. By 1935, the latest year for which Cleona Lewis compiled data on a basis comparable with earlier years, there were two important trends

Table 4.4: Per Cent Control by Major Powers in Key Producing Areas Outside the United States, 1929 and 1939

Year and Power	Producing Area		
	Western Hemisphere	Near East	Far East
1929			
United States	58		5
Britain and Holland	39	100	90
Other	3		5
1939			
United States	48	16	19
Britain and Holland	36	78	73
Other	16	6	8

Source: H. Williamson, R. Andreano, A. Daum and G. Klose, *The American Petroleum Industry, The Age of Energy, 1899-1959* (Northwestern University Press, Evanston, Ill., 1963), p.730.

evident in the geographical distribution of investment in production. First, there was a distinct shift from the Western to the Eastern Hemisphere, with the former's share declining from 98 to 85 per cent. Had Lewis compiled data for 1939, this shift would have been even more pronounced as a result of the nationalisations in Mexico and Bolivia. Within the Eastern Hemisphere, most of the investment was concentrated in the Netherlands East Indies. The Middle East had not yet come into full bloom.

Second, within the Western Hemisphere, there was a marked shift away from Mexico towards Venezuela. The latter had displaced Mexico as the prime outlet for US investment. Mexico's share, even before nationalisation, had fallen sharply, from more than three-fifths to less than one-fourth. These trends are shown in Table 4.5.

During the interwar period the pattern of petroleum investment changed on a functional as well as on a geographical basis. As was mentioned in chapter 3, the pre-World War I investments were essentially market-oriented. Following the war, they became supply-oriented, stemming from (1) the policy shift of the Rockefeller group from one of purchasing crude in the open market to the outright acquisition of reserves and (2) the adoption by the State Department of a policy encouraging US nationals to obtain reserves abroad. As can be seen in Table 4.6, almost three-fifths of US investments in petroleum in 1914 were in distribution, i.e. market-oriented. By 1935, the distribution share dropped to less than two-fifths. The control of

Table 4.5: Geographical Distribution of US Direct Foreign Investment in Oil Production[a], 1919 and 1935

Country	1919 Investment ($ millions)	(%)	1935 Investment ($ millions)	(%)
Western Hemisphere	321	98	742.5	85
Venezuela	(18)	(5)	(240)	(28)
Mexico	(200)	(61)	(206)	(24)
Colombia	(20)	(6)	(126)	(14)
Peru	(45)	(14)	(60)	(7)
Canada	(30)	(9)	(55)	(6)
Aruba			(45)	(5)
Trinidad	(5)	(2)	(7)	(1)
Central America	(3)	(1)	(3.5)	(b)
Eastern Hemisphere	7.5	2	130	15
Netherlands East Indies			(75)	(9)
Iraq, Palestine, Syria and Cyprus			(25)	(3)
Rumania	(7.5)	(2)	(20)	(2)
Arabia, including Bahrein			(10)	(1)
Totals	328.5	100	872.5	100

Notes: a. Includes refining. b. Less than one-half of one per cent.
Source: C. Lewis, *America's Stake in International Investment* (Brookings Institution, Washington, 1938), p.538.

Table 4.6: Functional Distribution of US Direct Foreign Investment in Oil, 1914, 1919 and 1935

	Total	Market-oriented (distribution)	Supply-oriented (production)[a]
1914			
Investment ($ millions)	343	200	143
(%)	100	58	42
1919			
Investment ($ millions)	604	275	329
(%)	100	45	55
1935			
Investment ($ millions)	1,381.5	509	872.5
(%)	100	37	63

Note: a. Includes refining.
Source: C. Lewis, *America's Stake in International Investment* (The Brookings Institution, Washington, 1938), pp.579, 588.

reserves became the crucial element in controlling both the international industry and the economies of other countries.

Along with the growing importance of production within total oil investments, oil became an interestingly important part of all investments. Between the years 1919 and 1935, as Table 4.7 shows, oil investments grew at a faster pace than total investments. This contrasts with the period between 1897 and 1914, when oil investments grew at a slower pace than total investments. Furthermore, in the pre-World War I period, oil ranked fifth in rate of growth while in the post-war period it ranked third.

Table 4.7: Per Cent Changes in US Direct Foreign Investment by Industry, 1919-1935

Industry	Change
Utilities	688
Manufacturing	135
Petroleum[a]	129
Other[b]	44
Mining and smelting	39
Sales organisation[c]	34
Agriculture and timber	0
Railroads	−12
Total	86

Notes: a. Includes exploration, production, refining and distribution. b. Mainly banks and insurance companies. c. Includes trading companies and sales branches and subsidiaries of large corporations. Excludes petroleum distribution.
Source: C. Lewis, *America's Stake in International Investment* (The Brookings Institution, Washington, 1938), p.605.

This increase in the relative importance of oil industry investment stemmed directly from the US Government policy of affording special treatment to oil investments abroad. Unlike in the pre-World War I era, when oil was treated as just another commodity, oil became the key commodity in US foreign economic policy. This interest in oil on the part of the government forged strong links between the industry and the State Department, links that were to grow even stronger in the post World War II years.

Notes

1. A. Sampson, *The Seven Sisters, The Greatest Oil Companies and The World They Shaped* (The Viking Press, New York, 1976), p.49.
2. Ibid.
3. H. O'Connor, *World Crisis in Oil* (Monthly Review Press, New York, 1962), p.280.
4. Ibid.
5. Sampson, *The Seven Sisters*, p.48.
6. Ibid.
7. Ibid., p.49.
8. Ibid., p.47.
9. Ibid., p.48.
10. O'Connor, *World Crisis in Oil*, p.281.
11. Sampson, *The Seven Sisters*, p.52.
12. O'Connor, *World Crisis in Oil*, p.53.
13. J. Stork, *Middle East Oil and the Energy Crisis* (Monthly Review Press, New York and London, 1975), pp.8-9.
14. C. Tugendhat and A. Hamilton, *Oil, The Biggest Business* (Eyre Methuen, London, 1975), p.66.
15. O'Connor, *World Crisis in Oil*, p.279.
16. Tugendhat and Hamilton, *Oil, The Biggest Business*, p.66.
17. Great Britain, *5 Parliamentary Debates* (Commons), LV (1913), pp.1475, 1477.
18. O'Connor, *World Crisis in Oil*, p.282.
19. Sampson, *The Seven Sisters*, p.55.
20. Tugendhat and Hamilton, *Oil, The Biggest Business*, p.68.
21. J. E. Hartshorn, *Oil Companies and Governments, An Account of the International Oil Industry in its Political Environment* (Faber & Faber, London, 1962), p.233.
22. P. de la Tramerye, *The World Struggle for Oil* (Knopf, New York, 1924), p.106.
23. Ibid., p.10.
24. US Federal Trade Commission, *The International Petroleum Cartel* (US Government Printing Office, Washington, 1952), p.54.
25. US Fuel Administration, *Final Report of the United States Fuel Administration, 1917-19* (Washington, US Government Printing Office, 1921), p.261.
26. Ibid., p.277.
27. Ibid., p.271.
28. Ibid., p.272.
29. In actuality the British, despite their misgivings about the mixed nationality of Shell, regarded Shell's reserves as part of Britain's reserves. In this instance their policy was exactly the same as that of the Americans. See Hartshorn, *Oil Companies and Governments*, p.233.
30. *Historical Statistics of the United States, Colonial Times to 1957: A Statistical Abstract Supplement* (US Government Printing Office, Washington, 1960), p.565.
31. S. Nearing and J. Freeman, *Dollar Diplomacy*, reprint (Monthly Review Press, New York and London, 1966), pp.85-9.
32. Ibid., pp.89-90.
33. C. Lewis, *America's Stake in International Investments* (The Brookings Institution, Washington, 1938), p.222.
34. D. Lawrence, *The True Story of Woodrow Wilson* (Doran, New York,

1924), p.100.

35. Nearing and Freeman, *Dollar Diplomacy*, pp.96-7.
36. Ibid., p.107.
37. Ibid., p.111.
38. Ibid., pp.111-12.
39. Lewis, *America's Stake in International Investment*, pp.221-2.
40. H. Williamson, R. Andreano, A. Daum and G. Klose, *The American Petroleum Industry, The Age of Energy, 1899-1959* (Northwestern University Press, Evanston, Ill., 1963), p.262.
41. Ibid., p.29.
42. M. Wilkins, *The Maturing of Multinational Enterprise: American Business Abroad from 1914 to 1970* (Harvard University Press, Cambridge, 1974), p.36.
43. In actuality Mexico moved into second place in 1918. This change in relative position was caused not only by the rapid expansion of Mexican production but also by the decline in Russian production following the Revolution and the Civil War.
44. Williamson, Andreano, Daum and Klose, *The American Petroleum Industry*, p.29.
45. H. Fraser, *Diplomatic Protection of American Petroleum Interests in Mesopotamia, Netherlands East Indies, and Mexico*. A Study Prepared for the Special Committee Investigating Petroleum Resources. US Senate. 79th Congress. 1st Session. Document No. 43 (US Government Printing Office, Washington, 1945), p.59.
46. Ibid., pp.72-3.
47. Williamson, Andreano, Daum and Klose, *The American Petroleum Industry*, p.731.
48. US Federal Trade Commission, *The International Petroleum Cartel*, pp.48-9.
49. Ibid., p.50.
50. Ibid., pp.50-1.
51. The companies were: Mexican Petroleum, controlled by Doheny; Texas; Gulf; Atlantic; Sinclair; Standard Oil (New York); and Standard Oil (New Jersey). See ibid., p.52.
52. Fraser, *Diplomatic Protection*, pp.3-4.
53. Ibid., p.3.
54. The Allies, with the exception of the United States, signed two peace treaties with Turkey. The first, the Treaty of Sevres, was signed in 1920 but was never put into effect because of the Kemal revolution in Turkey. The second, the Treaty of Lausanne, was signed in 1923. The US signed a treaty of amity and friendship with Turkey in 1923 but this treaty was not ratified by the Senate. See ibid., p.2.
55. Ibid., p.10.
56. Ibid., p.5.
57. Ibid.
58. US Federal Trade Commission, *The International Petroleum Cartel*, p.53.
59. Ibid., p.54. Of the seven companies listed in note 51 above, two — Sinclair and Texas — dropped out of the negotiations. The interests of Mexican Petroleum were taken over by Pan American.
60. Ibid. In 1933 Standard Oil (New York) merged with the Vacuum Oil Company and changed its name to the Socony-Vacuum Oil Company.
61. Ibid., p.53.
62. Ibid., pp.65-7.
63. Fraser, *Diplomatic Protection*, p.32.
64. Ibid., p.33.

65. Ibid.
66. Ibid., p.34.
67. Ibid., p.35.
68. Ibid., p.38.
69. Ibid., p.41. The note was dated April 12, 1921.
70. Ibid., p.42.
71. Ibid., pp.45-6.
72. Ibid., p.48.
73. Ibid., p.51.
74. Ibid.
75. Luis Alcala Sucre, 'The Impact of Multinational Oil Companies on Venezuela: The Oil Companies in General and Gulf Oil in Particular' in J. P. Gunneman (ed.), *The Nation-State and Transnational Corporations in Conflict with Special Reference to Latin America* (Praeger, New York, Washington and London, 1975), pp.177-223.
76. Ibid., p.182.
77. O'Connor, *World Crisis in Oil*, pp.135-6.
78. Neil Jacoby, *Multinational Oil, A Study in Industrial Dynamics* (Macmillan, New York, 1974), p.32.
79. Luis Vallenilla, *Oil: The Making of a New Economic Order, Venezuelan Oil and OPEC* (McGraw-Hill, New York, 1975), p.7.
80. O'Connor, *World Crisis in Oil*, p.131.
81. Vallenilla, *Oil: The making of a New Economic Order*, p.7.
82. Ibid.
83. O'Connor, *World Crisis in Oil*, p.131.
84. Ibid., pp.129-30.
85. Tugendhat and Hamilton, *Oil, the Biggest Business*, p.79.
86. O'Connor, *World Crisis in Oil*, p.130.
87. Ibid., p.132.
88. Ibid., p.131.
89. Ibid.
90. Wilkins, *The Maturing of Multinational Enterprise*, p.115.
91. Ibid.
92. Vallenilla, *Oil: The Making of a New Economic Order*, p.36.
93. O'Connor, *World Crisis in Oil*, p.131.
94. Wilkins, *The Maturing of Multinational Enterprise*, p.115.
95. O'Connor, *World Crisis in Oil*, p.136.
96. Ibid.
97. Vallenilla, *Oil: The Making of a New Economic Order*, pp.30-1.
98. Wilkins, *The Maturing of Multinational Enterprise*, p.115.
99. Vallenilla, *Oil: The Making of a New Economic Order*, p.29.
100. Williamson, Andreano, Daum and Klose, *The American Petroleum Industry*, p.541.
101. E. H. Shaffer, *The Oil Import Program of the United States, An Evaluation* (Praeger, New York, Washington and London, 1968), p.9.
102. Wilkins, *The Maturing of Multinational Enterprise*, p.209.
103. Ibid., p.210.
104. O'Connor, *World Crisis in Oil*, p.137.
105. P. Collier and D. Horowitz, *The Rockefellers: An American Dynasty* (New American Library, New York, 1976), pp.207-8.
106. Sampson, *The Seven Sisters*, p.78.
107. Collier and Horowitz, *The Rockefellers*, p.208.
108. Vallenilla, *Oil: The Making of a New Economic Order*, p.53.
109. Ibid., pp.55-6.

110. Ibid., pp.56-7.
111. Ibid., p.57.
112. O'Connor, *World Crisis in Oil*, p.113.
113. Vallenilla, *Oil: The Making of a New Economic Order*, p.54.
114. Ibid., pp.57-8.
115. Jacoby, *Multinational Oil*, p.41.
116. US Federal Trade Commission, *The International Petroleum Cartel*, p.130.
117. Ibid., p.71.
118. Gulf sold its shares in IPC in 1931 to free itself from the 'Red Line' agreement.
119. Ibid., pp.130-1.
120. N. Robertson (ed.), *Origins of the Saudi Arabian Oil Empire, Secret U.S. Documents, 1923-1944* (Documentary Publications, Salisbury, North Carolina, 1979), p.i.
121. Ibid., p.19.
122. Ibid., p.21.
123. Ibid.
124. Ibid., p.1.
125. Ibid., p.25.
126. Ibid., pp.52-3.
127. Ibid., p.55.
128. Ibid., p.48.
129. Ibid., pp.49-50.
130. Ibid., p.84.
131. Ibid., pp.83-4.
132. Ibid., p.88.
133. Ibid., p.97.
134. US Federal Trade Commission, *The International Petroleum Cartel*, p.116.
135. Ibid., pp.114-16.

5 THE RELATIONSHIP BETWEEN THE OIL INDUSTRY AND THE AMERICAN STATE — POST WORLD WAR II ERA

World War II illustrated even more dramatically than World War I the importance of oil in modern warfare. Both the Allied and Axis forces were highly mechanised with large fleets of aircraft, lorries, tanks, self-propelled artillery and ships. All this equipment consumed enormous quantities of oil.

Many key battles were fought over oil. The German drive to Stalingrad was motivated partly by the desire to cut off oil shipments to the Red Army coming up the Volga from Iran and partly by an attempt to secure the oil-rich Caucasus for the Third Reich. The Japanese invasion of Indonesia was likewise prompted by the need to obtain that area's oil. The massive Allied air raids on Ploesti in Rumania were aimed at depriving Germany of its major oil source. These are but a few of the many examples of oil's strategic role during the war.

Within the United States oil policy was a matter of paramount concern. As in World War I, the US supplied the Allies with most of their oil. The US furnished nearly 5 billion of the 7 billion barrels consumed by the Allies between 1941 and 1945.[1] Also, as in World War I, the US feared that it would soon deplete its own oil resources. As an example of this concern, William C. Bullitt, Under Secretary of the Navy, drafted a letter to President Roosevelt in June 1943, which stated:

> The crude petroleum reserves of the United States, at either present rates of withdrawal or peace-time rates, will be totally exhausted in about 14 years — unless new reserves of major importance should be discovered. Our leading petroleum geologists express the opinion that there are no known major structures the development of which would make more than a minor change in the situation.[2] (Italics as in original)

Bullitt then raised the question of finding secure future oil supplies. Following Bullitt's letter, top government leaders engaged in an extensive discussion of postwar oil policy. Before analysing in detail the evolution of this policy, it is first necessary to examine the environment in which this policy was formulated. As a result of the war, the world power structure had been drastically changed. The United States became the dominant power in the world.

Dominance of the United States

The United States, unlike most of the belligerents, escaped the ravages of war. Together with its northern neighbour, Canada, it emerged from the struggle with an enhanced economic base. GNP (Gross National Product), for instance, grew in real terms by 49 per cent between 1939 and 1946. Business investment, the key to increased capacity, expanded much more — by 88 per cent.[3] Furthermore, US casualties were relatively light, both in comparison to the other belligerents, including Canada, and to US casualties in previous wars. Though the US suffered over one million casualties, including more than 400,000 deaths, the ratios of both casualties and deaths to armed forces personnel were about equal to the ratios in World War I, approximately 7 per cent for total casualties and 2.5 per cent for deaths.[4] The United States, however, was engaged in the second war for a far longer period (46 months) than in the first one (19 months). Thus the comparative losses of the United States were smaller in the second conflict. In only one other conflict, the US Civil War, did the US forces fight for as long a time as in World War II. In that conflict the ratios of total casualties and deaths to the armed forces personnel were considerably higher, about 29 per cent for casualties and 17 per cent for deaths.[5]

Furthermore, since the US was neither invaded nor bombed, it suffered few civilian casualties. At the end of the war the US thus had an intact military machine, an effective civilian labour force and an enlarged industrial base. In addition, it had acquired a significant lead in technology. It developed and perfected many new weapons, including the atom bomb, on which it had a monopoly. Given its superb industrial and military machine and its possession of the bomb, the US was in a position to impose a *Pax Americana* on most of the world.

The only parts of the world in which the US was unable to assume a dominant position were Eastern Europe, the Soviet Union and the Communist-held areas of China. Despite the immense devastation these lands suffered, they still managed to retain an effective military capacity. The Soviet Union, in particular, emerged from the war as a major military power. Though relatively weak in comparison to the US, it was nevertheless strong enough to keep the US from dominating these areas.

The refusal of the Soviets to accept a *Pax Americana* was one of the factors that led to the Cold War. A detailed analysis of the origins of the Cold War is beyond the scope of this book.[6] Nevertheless it is useful to question the conventional wisdom that the Cold War represented a defensive American response to the threat of Soviet expansionism. American policy in that period was anything but defensive. It followed the traditional policy of acquisition without annexation, i.e. the policy of Carl Schurz. Under the guise of defending the 'Free World', the US was able to exercise a dominant influence over most non-Communist nations.

Acquisition without Annexation

The US acquired military bases throughout the world. Schurz, it should be recalled, advocated such a policy as a means of obtaining economic dominance. By 1968 the US had bases and facilities in 64 different countries, as can be seen in Table 5.1 Magdoff has noted that this was far more than in any previous period in US history, including World War II. The 'U.S. armed forces in the 1920's', Magdoff pointed out,

Table 5.1: Geographical Distribution of US Bases and Facilities, 1968

Region	Number of countries in which US armed forces were stationed
Latin America	19
Europe	13
Africa	11
Near East and South Asia	11
East Asia (including Australia)	10
Total Countries	64

Source: H. Magdoff, *The Age of Imperialism, The Economics of U.S. Foreign Policy* (Modern Reader Paperbacks, New York and London, 1969), p.42.

'were stationed in only three countries abroad. During World War II, U.S. armed forces were to be found in 39 countries.'[7] Another interesting aspect is that a very large number of facilities were located in areas like Latin America, Africa, South Asia and East Asia to which the Soviet Union could not have possibly posed a serious military threat. But the presence of these facilities did create a favourable climate for US trade and investment. American forces in an area presumably ensured political stability, a necessary ingredient of any large-scale international investment programme.

The Export of Capital

One of the cornerstones of US foreign economic policy was the
encouragement of private investment abroad. This has been a consistent
aim of all post World War II US governments, irrespective of party.
President Truman, for instance, stated in his 1949 inaugural address
that 'we should foster capital investment in areas needing development'.[8]
President Eisenhower in 1953 said that 'a serious and explicit purpose
of our foreign policy [is] the encouragement of a hospitable climate
for investment in foreign nations.'[9] Dean Rusk, Secretary of State
under Presidents Kennedy and Johnson, told a Congressional
committee hearing that foreign countries should:

> create conditions which will be attractive to the international
> investor, the private investor. So our influence is used wherever it
> can be and persistently, through our Embassies on a day-to-day basis,
> in our aid discussion and in direct aid negotiation, to underline the
> importance of private investment.[10]

Since the bulk of private foreign investment then, and now, came from
the United States, Dean Rusk was really talking about American
private investment. His remarks, like those of Presidents Truman and
Eisenhower, were essentially no different than those of President Taft
in his famous 'Dollar Diplomacy' speech.

Dean Rusk moreover mentioned an additional policy tool, the tool
of foreign aid. In Taft's day, the US, which was still in a net debtor
position, was not able to dispense large sums of money for the purpose
of opening up trade and investment opportunities for its corporations.
In contrast, by the end of World War II the US had become the world's
leading creditor and was thus in a position to use aid as a key policy
tool. As President Kennedy stated in 1962, 'foreign aid is a method
by which the United States maintains a position of influence and
control around the world . . .'[11]

By pouring dollars into a dollar short world, the US aid programme
created markets for its exports and outlets for its investment capital.
In the early post-war years the aid outlays enabled the US to amass
huge balance-of-trade[12] surpluses. As Table 5.2 shows, aid expenditures
in the first ten post-war years, the period in which the dollar shortage
was most severe, equalled $53 billion, a sum only $3 billion less than
the trade surplus. Had this supply of dollars not been forthcoming, the

Table 5.2: Relationship between US Foreign Aid Expenditures and the Balance-of-Trade Surplus, 1946-1955

	(1) Net aid expenditures[a] ($ billions)	(2) Balance-of- trade surplus[b] ($ billions)	(3) Relationship (1)/(2) x 100 (%)
1946	5.4	6.7	81
1947	6.2	10.1	61
1948	5.3	6.0	88
1949	5.8	5.4	107
1950	4.1	1.5	273
1951	4.6	4.3	107
1952	5.0	5.1	98
1953	6.3	5.5	114
1954	5.0	5.8	86
1955	4.8	5.4	89
Totals 1946-55	52.5	55.8	94

Notes: a. Includes civilian and military grants and credits less loan repayments.
b. Merchandise exports less merchandise imports. Includes exports made under military grants programmes.
Source: US Department of Commerce, *Balance of Payments Statistical Supplement*, rev. edn. (US Government Printing Office, 1963), pp.3, 154-7.

dollar hungry world would not have been able to purchase the goods produced by American corporations and the US could not have acquired such huge surpluses.

The transfer of dollars also created a favourable investment climate; US corporations investing abroad were certain that they would receive dollars for their products sold abroad. In addition this aid was often used to provide the infra-structure necessary to induce private investment. Furthermore, specific aid programmes, like Public Laws 480 and 128, allowed American firms to borrow the foreign currencies the US had accumulated in less developed countries to make investments in those countries. The government also provided investment guarantees generally covering losses up to 75 per cent of the investment. Finally Congress in 1962 passed the Hickenlooper Amendment to the Foreign Assistance Act, which provided for mandatory cut-offs of aid to any country expropriating US-owned property unless that country made immediate and full payment to its owners. Perhaps more than any other piece of legislation, the Hickenlooper Amendment bared the link between foreign aid and investment.

The International Monetary Regime

The international monetary regime also facilitated the flow of capital abroad. Under the rules of the IMF (International Monetary Fund), adopted at the insistence of the United States at the Bretton Woods Conference of 1944, the US dollar became, along with gold, the means of settling international accounts. This special status of the dollar removed normal restraints to US international investments. American firms seeking assets abroad did not have to acquire either gold or foreign currency to purchase them. They purchased them with US dollars. Because of the wide-spread acceptability of the dollar, they were able to continue expanding abroad even after the United States began incurring large balance-of-payments deficits in the 1950s. As long as foreign governments and banks were willing to hold dollar reserves, American companies were free to invest wherever they pleased. Foreigners, on the other hand, had to worry about the costs of acquiring either US dollars or gold before they could invest abroad. Thus, as a result of both the US aid programmes and the status of the dollar, US direct foreign investment grew at a rapid rate.

As can be seen in Table 5.3 the book value of foreign investment rose at a faster rate than several other key indicators in each of the three post-war decades with only one exception. That exception occurred in the third decade when merchandise exports expanded at a faster pace than foreign investment. During this decade the US dollar lost much of its status as an international currency. Because of the persistent balance-of-payments deficit, the dollar shortage of the 1940s and the early 1950s had turned into a dollar glut. By the late 1960s many governments and banks refused to add to their dollar reserves. Instead of dollars they demanded gold or other hard currencies. This loss of acceptability forced a devaluation of the dollar. As a consequence, foreign investments became more costly to American firms. At the same time American goods became cheaper to foreigners. Thus merchandise exports grew at a faster rate than foreign investments. Nevertheless the growth in foreign investments was significantly higher than in other key indicators, such as GNP, Net Private Domestic Investment and Net Corporate Investment.

There is little doubt that the export of capital played a major role in the American economy in the post-war years. Not only did foreign investment expand much faster than domestic investment but also the earnings from foreign investment became a key element in the United States balance-of-payments position. As Table 5.4 indicates, income from foreign investments became an increasingly larger portion

Table 5.3: Per Cent Changes in Book Value of Private Direct Foreign Investment, Gross National Product, Net Private Domestic Investment, Net Private Domestic Investment of the Corporate Sector, and Merchandise Exports, 1946-1955, 1955-1965 and 1965-1975

	1946-55	1955-65	1965-75
Foreign Investment	168	155	152
GNP	91	72	122
Domestic Investment	96	65	−47
Corporate Investment[a]	103	112	38
Exports	22	84	304

Note: a. Net investment of the corporate sector excludes residential and agricultural investments. The latter are included in domestic investment, which encompasses all investment in the economy.
Sources: *Historical Statistics of the United States, Colonial Times to 1957: A Statistical Abstract Supplement* (US Government Printing Office, Washington, 1960), p.565; M. Wilkins, *The Maturing of Multinational Enterprise: American Business Abroad from 1914 to 1970* (Harvard University Press, Cambridge, Mass., 1974), p.329; *Economic Report of the President Transmitted to the Congress January 1980* (US Government Printing Office, Washington, 1980), pp.203, 216, 219, 222, 316 and 321; *Economic Report of the President Transmitted to the Congress January 1979* (US Government Printing Office, Washington, 1979), p.298.

Table 5.4: Relationship between Net Income on Foreign Direct Investment and the Balance-on-Goods-and-Services, 1946-1978

	(1) Income[a] ($ billions) (annual average)	(2) Balance[b] ($ billions) (annual average)	(3) Relationship (1)/(2) x 100 (%)
1946-9	1.3	8.2	16
1950-4	2.3	2.7	85
1955-9	3.5	3.9	90
1960-4	5.0	6.8	74
1965-9	7.1	5.4	132
1970-4	12.8	5.3	242
1975-8	27.7	3.7	749

Notes: a. Investment income payments to Americans by foreigners minus similar payments by Americans to foreigners. Included in investment income are interest, dividends, royalties and service fees. The latter two are classified by the US Department of Commerce as 'other service income'. In this table, they are included in investment income because they are related to the technology transfer process which normally accompanies international investment. b. The balance-on-goods-and-services encompasses the trade balance plus the balance on services, like insurance, travel, freight and investment income.
Source: *Economic Report of the President Transmitted to the Congress January 1980* (US Government Printing Office, Washington, 1980), p.316.

of the balance-on-goods-and-services[13] and has actually exceeded the balance since the mid-1960s. The increase in investment income partially offset the decline in the trade balance, which during the 1970s turned from a surplus to a deficit. Thus, by the 1970s the American Empire reached the same position as the British Empire at the turn of the century, when Britain became a net importer of merchandise and relied on income from foreign investments to balance its international accounts. The United States has similarly become dependent on the income from its overseas investments.

US Trade Balance

As Table 5.5 shows, the United States's balance-of-trade surplus reached a peak in the early post-war years and has followed a secular downtrend ever since. By the 1970s the balance-of-trade surplus turned into a deficit. The US deficit differed however from the British in one significant respect. While the British deficit was caused by the imports of agricultural goods and raw materials, the American one was caused by the importation of manufactured goods. A portion of these manufactured goods was produced by the overseas facilities of American corporations. Thus, in the development of the American Empire the export of capital first tended to replace the export of goods and later to displace the domestic production of goods.

Table 5.5: United States Balance-of-Trade, 1946-1978

Years	Balance (annual average) ($ billions)
1946-9	7.0
1950-4	2.9
1955-9	3.7
1960-4	5.4
1965-9	2.8
1970-4	−2.1
1975-9	−16.3

Source: *Economic Report of the President Transmitted to the Congress January 1980* (US Government Printing Office, Washington, 1980), p.316.

It is significant that the US balance-of-trade deficit was not caused by the rise in oil prices. The first deficit occurred in 1971, a full three years before the four-fold oil price increase of 1974. In 1971 manufactured imports accounted for approximately two-thirds of the import bill.[14] The oil bill was about 8 per cent.[15] By 1978, when oil accounted for almost one-fourth of all imports, manufactured imports still accounted for more than one-half.[16]

The rise in imports from US owned facilities abroad was directly attributable to the growth in investment in manufacturing. As can be seen in Table 5.6 more than 40 per cent of all direct investment in 1970 was in manufacturing. Manufacturing, in comparison, accounted for only one-third of investment at the war's end.[17] This predominance of manufacturing investment stands in contrast to earlier years, when manufacturing ranked well below mining and smelting (see Table 2.4).

Table 5.6: Distribution of US Direct Private Investments by Industry, 1970

Industry	Investment ($ billion)	(%)	Per cent change (1946-1970)
Manufacturing	32.3	41	1,246
Petroleum[a]	21.7	28	1,450
Trade	6.6	8	1,000
Mining and smelting	6.2	8	675
Public utilities[b]	2.9	4	123
Other[c]	8.5	11	1,114
Totals	78.2	100	

Notes: a. Includes production, refining, marketing, transportation and shipping. b. Includes transportation. c. Includes agriculture, sales organisations, banks and insurance companies.
Source: M. Wilkins, *The Maturing of Multinational Enterprise: American Business Abroad from 1914 to 1970* (Harvard University Press, Cambridge, Mass., 1974), p.329.

As in the past, most manufacturing investment was concentrated in the developed countries. The data in Table 5.7 show that almost three-fourths of manufacturing investment was located in Europe and Canada. Only three-fifths of total investment was placed in these areas. This pattern was similar to the one prevailing in 1914 (see Tables 2.5 and 2.6). The typically colonial pattern had not changed.

What had changed was the increasing influence the US corporations exercised over the developed countries. In many countries, and especially in Canada, US corporations controlled key sectors of the economy. By the end of the 1960s the American Empire had expanded beyond the wildest dreams of Carl Schurz. American enterprise truly ruled the world.

Table 5.7: Geographical Distribution of US Direct Private Foreign Investment, 1970

Region	Total ($ billion)	Total (%)	Manufacturing ($ billion)	Manufacturing (%)	Petroleum[a] ($ billion)	Petroleum[a] (%)
Europe	24.5	31	13.7	42	5.5	25
Canada	22.8	29	10.1	31	4.8	22
Latin America	14.8	19	4.6	14	3.9	18
Asia	5.6	7	1.5	5	3.0	14
Africa	3.5	4	0.5	2	2.1	10
Oceania	3.5	4	1.8	6	0.7	3
Unallocated	3.5	4	0.1	b	1.7	8
Totals	78.2	100[c]	32.3	100	21.7	100

Notes: a. Includes production, refining, marketing, transportation and shipping. b. Less than one-half of one per cent. c. Because of rounding, the total may not equal the sum of the column.
Source: M. Wilkins, *The Maturing of Multinational Enterprise: American Business Abroad from 1914 to 1970* (Harvard University Press, Cambridge, Mass., 1974), p.330.

Oil in America's Expansion

The oil industry played a key role in this expansion. Though investment in manufacturing in 1970 was greater than in oil, the rate of growth in oil investments during the post-war period was, as Table 5.6 shows, greater than for any other industry. As has already been indicated in Tables 3.5 and 4.7, oil's growth rate in earlier periods was below that of other industries. The rapid expansion of oil investments in the post-war era was directly related to the need to consolidate and secure the growing American Empire.

Bullitt's Proposals

It was against this background that William Bullitt drafted his 1943 warning to President Roosevelt. He, like most other US leaders, was aware of the new role America was to play in the post-war world. Perturbed by the prospect that an oil-deficient United States might not be able to play this role, Bullitt told the President:

> we can and must act to provide additional reserves outside the continental limits of the United States. Oil is and will remain for some decades a vital element both for our national defense and our national economy. *To acquire petroleum reserves outside our*

boundaries has become, therefore, a vital interest of the United States.[18] (Italics as in original)

What Bullitt proposed was the establishment of a government-owned Petroleum Reserves Corporation which would 'acquire for the Government of the United States a controlling interest in proved foreign oil reserves'.[19] As a first step he suggested that the Corporation 'give immediate attention to the acquisition of a controlling interest in the concessions now held in Saudi Arabia by the California Arabian Standard Oil Company',[20] which was then owned jointly by Standard Oil of California and the Texas Company.

Interestingly enough, Bullitt drew a parallel between his proposed Corporation and the British Government's acquisition of stock in Anglo-Iranian. He argued that if the US Government did acquire the controlling interest in the Saudi Arabian company and rendered to it 'support of the same effective nature that the British Government has given and is giving the Anglo-Iranian Oil Company, the controlling interest in which is owned by the British Government — it might be possible to retain the concession in American hands and develop it successfully'.[21]

He emphasised why he favoured this British approach rather than the American one of giving diplomatic backing to private US companies operating abroad. In rebutting a suggestion that the government, instead of purchasing stock in the Saudi company, buy an interest in reserves of about 300 million barrels, he stated:

> To accept any such proposal would be to relapse into the 'dollar diplomacy' of a dead era. The Government of the United States would be undertaking to back up with all its diplomatic and other resources private capitalists whose concessions will be relatively valueless without government support. In exchange for the reservation of 300,000,000 barrels of oil, our Government would be virtually guaranteeing private owners in the possession of 20 billion barrels of oil. *No such policy should be countenanced by our Government.*[22] (Italics added)

Finally Bullitt urged that the Corporation should not limit itself to Saudi Arabia. He wanted it to 'explore all other possibilities in all quarters of the earth'.[23] Given only a 14 years' supply in the United States, he argued that with the acquisition of foreign oil reserves the US 'domestic reserve should be kept in the ground as a strategic

reserve and that our current needs should be filled from sources
of oil outside the continental limits of the United States'.[24]   Thus,
while ostensibly denouncing 'dollar diplomacy', he advocated a
policy which would denude the world of its oil in order to serve
the interests of the United States. The world's oil henceforth was
to belong directly to Uncle Sam. Oil imperialism was to replace dollar
diplomacy.

While the US Government gave serious consideration to Bullitt's
import proposal, it was never able to implement it. The reason lay in
the private ownership of domestic oil resources. Those companies with
large domestic reserves were unwilling to forego income for long periods
of time in order to permit imports to serve the US market. The loss of
present income for some uncertain future income would have reduced
the present values of their reserves. No private company would be
willing to incur such a loss unless it received compensation in return.
The government, quite understandably, was unwilling to pay the
private producers the large sums necessary to induce them to shut-in
their wells. The only other alternative would have been nationalisation.
That was both costly and politically unacceptable.

In retrospect, Bullitt's proposal was a sensible one for the United
States. Had United States wells been shut-in, Middle Eastern and other
foreign oil would have been depleted at a much faster rate. The United
States and its allies furthermore would not have been vulnerable to
any embargo. The untapped reserves of the United States would have
been able to offset any severe shortages. OPEC would probably never
have been formed. If it had, its bargaining power would have been
seriously weakened as long as the United States was willing to throw
its reserves on the market to prevent prices from rising. But in a
capitalist state, where private interests must prevail, it is often difficult,
if not impossible, to act in the national interest, especially if the
national interest seriously conflicts with the interests of powerful
private parties.

Petroleum Reserves Corporation

While Bullitt's import proposal was not adopted, his suggestion for a
Petroleum Reserves Corporation was. The Petroleum Reserves
Corporation was created by Executive Order, i.e. by the Executive
Branch and not by Congress, in July 1943. Its Board of Directors
consisted of the Secretary of State, the Secretary of War, the Secretary
of the Navy, the Secretary of the Interior and the Director of the
Office of Economic Warfare.[25] The Corporation was obviously a top-

level body run by some of the most powerful men in the government.

Despite protracted negotiations the Corporation failed to acquire control of the Saudi company. Initially the two partners agreed to let the Corporation buy one-third of the shares. Later the Texas Company changed its mind, claiming that the Corporation did not offer it enough for the shares.[26] Eventually the scheme was abandoned and the corporation was dissolved in 1944.[27]

It is interesting to note that this high-level Corporation was unable to force either oil company to sell its shares to it in order to protect the presumed national interest. This incident illustrates what the true power relationships in the United States were during the war. It also shows how well-entrenched the oil industry had become in the American Establishment since the dissolution of the Standard Oil Company in 1911.

The Industry's Alternative

The entire oil industry opposed both the establishment of the Corporation and the acquisition by that Corporation of stock in private companies. The Petroleum Industry War Council, an organisation representing virtually every oil company in the United States, adopted a resolution stating that 'oil development can best be handled by private enterprise'.[28] It went on to assert that 'oil in the hands of nationals of the United States is equally available for national security with oil owned or financially shared in by the Government of the United States'.[29] It then concluded that the 'Government of the United States should encourage private American enterprise to engage in the development of oil resources abroad'.[30]

From these general statements it made some specific points. For instance, it gave the following estimate of how much foreign reserves the US companies should acquire.

> The combined domestic and foreign oil reserves held by nationals of the United States constitute a smaller proportion of the petroleum reserves of the world than the ratio of United States consumption of petroleum to total world consumption. Action is needed to enlarge the reserves under the stewardship of nationals of the United States . . .[31]

This doctrine, if carried to its logical conclusion, would have given the United States control of most of the world's mineral resources since the United States was then, and still is, the largest single consumer of most raw materials in the world. Presumably mere consumption, no matter how wasteful, justifies the concomitant reserves. Such were the benefits that could be expected from a *Pax Americana*.

The resolution also stressed the need for acquiring many types of facilities throughout the world.

> To enhance national security many and widely distributed sources of products for military supply should be in the hands of United States nationals. Such sources have been developed by private industry and in many areas can only be so developed. Only through distributing facilities as well as sources of supply can these requirements be assured.[32]

This paragraph, it should be noted, speaks of investment in general, not merely oil investments. It is interesting in that it reveals the attitude of an important segment of American business towards the link between economic and military power. It is a relationship of interdependence. Military power creates the conditions necessary for economic expansion. Economic power provides the basis for military power. Multinational corporations were thus to be military as well as economic entities.

This link between the economic and military variables is implied in the following statement:

> The government of the United States can serve the interests of a greater number of its own nationals and at the same time promote the interests of foreign countries by means of a foreign oil policy designed to reduce the political risks inherent in foreign oil operations.[33]

While neither 'political risks' nor the designs to reduce them are explicitly defined, one can presume that the most worrisome risk is nationalisation and that military intervention can reduce this risk. It is significant that nowhere in the resolution is military intervention explicitly ruled out as a policy instrument for reducing political risks.

Collado's Proposals

Once the Petroleum Reserves Corporation had been scuttled the United States Government had no choice but to adopt the main policy lines of the Petroleum Industry War Council. Though State Department officials expressed some reservations, they were forced to accept it. In a memorandum dated 29 March 1945 to William Clayton, head of

the State Department's Economic Affairs Division and Assistant
Secretary of State, Emilio Collado of the Department's Office of Fuel
Development questioned whether national security required that
Saudi Arabian oil be controlled by American companies. Collado
wrote:

> It is not of vital importance to the military security of the United
> States that Arabian oil remain in the hands of an American
> company. The Navy has admitted that Arabian oil would not be
> subject to American strategic control in the next war. The British
> would be our Allies, and if it could be controlled by them, it
> would be as good as ours.[34]

There are two points of interest in this statement. One is Collado's
acceptance of the inevitability of another war even before World War II
had come to its end. The other is the willingness to let the British have
custody of Arabian oil. The latter point represented a break with the
traditional State Department's view of Britain. Recognising this,
Collado commented:

> Most foreign service officers who have served in the Middle East
> have an inferiority complex vis-à-vis the British, resulting from
> years of being out-smarted and operating without much interest
> or backing from home. Arabian oil is the greatest opportunity
> they have ever had to stimulate interest in the Middle East and
> U.S. government backing against their long-time adversaries, the
> British. They will not acknowledge that cat and dog fights with
> the British in the Middle East back yard have now become foolish
> and futile, since the British can be handled quite simply by a
> firm hand in Washington.[35]

Collado, who was later to leave the State Department to become an
executive in Standard Oil (N.J.), thus explicitly recognised the
change in international power relations brought about by the war.
Britain had been reduced to a dependency of the United States.

Collado further argued that there were distinct advantages in
giving Saudi oil to a British firm. Among them would be the ability
to implement a modified version of Bullitt's import plan by having
Middle East oil replace Western Hemisphere oil in the Euroepan
market.

Arabian oil, if properly handled, could contribute to the military security of the western hemisphere. If developed and marketed in Europe, replacing western oil that might be marketed there, western reserves might be conserved to the extent of such replacement. The unpopularity of this procedure with the great portion of the American petroleum industry not represented in the Middle East will make it very difficult for the U.S. Government to pursue an aggressive course in this regard. It is interesting to contemplate that the desired result would more likely be obtained if Arabian oil were in British hands, so that the replacement process in Europe would take place through British competition without any encouragement from the U.S. Government. The U.S. Government could also avoid embarrassment if the American oil company itself carried out the replacement without undue assistance from the Government (though the progress of an American company will be slow relative to that of a British Company because of the dollar payment difficulty).[36]

Despite these advantages, Collado recognised that there would be political difficulties in implementing his proposal. In an earlier memorandum to Clayton, he noted that 'it is probable that the State Department would be subjected to severe public criticism if the concession were lost either to Britain or Russia'.[37] Thus, despite Collado's misgivings, Saudi oil remained firmly in the hands of the private US companies. Given the government's view of the essential nature of oil, it had no choice but to back these companies in their relations with foreign governments.

Expansion in the Middle East

After rejecting Britain's surrogate role, American policy quite logically concentrated on replacing Britain as the Middle East's dominant oil power. As Table 5.8 shows the US at that time was the second oil power in the Middle East, controlling only 40 per cent of the fixed assets there. Britain, despite its weakened condition, retained 44 per cent of the assets. By 1959, the Americans owned 50 per cent, the British, 18 per cent. A number of factors contributed to this displacement of Britain.

Saudi Arabia. The first was the strengthening of the American position in Saudi Arabia. During the war King Ibn Saud became conscious of his country's vast oil potential. Eager to increase his revenues the King pressed ARAMCO to increase production. In response to his

Table 5.8: Estimated Distribution of Gross and Fixed Assets in the Petroleum Industry of the Middle East, by National Ownership and Control, Beginning of 1947 and 1959 (per cent)

Country	1947	1959
United States	40	50
United Kingdom[a]	44	18
France	8	5
Netherlands[a]	8	3
Others[b]		24
Total	100	100

Notes: a. Shell's investments were allocated as follows: 60 per cent to the Netherlands and 40 per cent to the United Kingdom. b. This mainly represents the holdings of the National Iranian Oil Company. As part of the settlement following the restoration of the Shah, the Iranian Government in 1954 took over the properties held by the Anglo-Iranian Oil Company. It also signed an agreement with the Consortium, known as Iranian Oil Participants Ltd, giving that group production and exploration rights in southern Iran and the right to operate the refinery at Abadan. The national distribution of ownership of the Consortium was: United Kingdom, 46 per cent; United States, 40 per cent; Netherlands, 8 per cent and France, 6 per cent.
Source: C. Issawi and M. Yeganeh, *The Economics of Middle Eastern Oil* (Praeger, New York, 1962), pp.59, 175-6.

request the owners of ARAMCO, Standard Oil of California and the Texas Company, laid plans to build a pipeline across the country to the Eastern Mediterranean. This pipeline required new markets in Europe, an area in which these companies had only limited marketing facilities. Both the companies and the US Government feared that ARAMCO might lose its concession if it could not market this oil.

At this point the State Department intervened. It wanted to hold on to what Walter W. Birge, Jr, American Vice-Consul in Dhahran, Saudi Arabia described as 'probably the largest oil field in the world — America's new frontier'.[38] In 1945 the State Department suggested to ARAMCO that it make an arrangement with the American participants in IPC to dispose of its oil. The Department proposed that it either sign a marketing agreement with these companies, Standard Oil (New Jersey) and Socony-Vacuum, or have them become partners in ARAMCO.[39] After protracted negotiations, Standard and Socony bought into ARAMCO in early 1947.[40] With the entry of these two partners, the ownership of ARAMCO was distributed as follows: Standard Oil in California, 30 per cent; the Texas Company, 30 per cent; Standard Oil (New Jersey), 30 per cent; and Socony-Vacuum, 10 per cent.[41]

The agreement was advantageous to all sides. The original partners could now go ahead with plans for expansion, knowing that the new partners would find markets for the increased crude production. In addition, Jersey and Socony paid California and Texas $125 million, an amount nearly equal to their investment in Saudi Arabia.[42] Thus the original partners were able to recoup their investment and still retain 60 per cent of the concession. Both Jersey and Socony, for their part, needed increased crude supplies for their growing markets. Jersey, in addition, wanted to make up for the loss of its Rumanian supplies following the nationalisation of its properties in that country. Before the war Jersey had supplied much of its European markets from Rumania.

That the Saudi venture, involving a joint production and marketing arrangement between four of the US's largest oil companies, may have violated the anti-trust laws deterred neither the companies nor the State Department. These laws apparently were not to stand in the way of the national interest. Nevertheless, the companies did find it convenient to invoke the anti-trust laws when these laws corresponded to their interests. The entry of Jersey and Socony into Saudi Arabia broke the 'Red Line' agreement. That agreement, it should be recalled, prohibited any of the IPC partners from obtaining concessions in the territories of the former Ottoman Empire without prior permission of the other partners. In actuality, the IPC as a whole was supposed to obtain any new concession. Since at least two of the partners, CFP and the Gulbenkian interests, objected to the move by Jersey and Socony, the latter two companies notified the other partners that they considered that the 'Red Line' agreement was no longer binding upon them.[43] The agreement, they claimed, was 'in restraint of trade and contrary to public policy and void and unenforceable in law'.[44] The US Federal Trade Commission wryly commented:

> Thus, Standard and Socony followed the red-line agreement for almost 20 years without making any serious objections, but when they wished to obtain an interest in Aramco, they described the agreement as one 'in restraint of trade and contrary to public policy and void and unenforceable in law'.[45]

By making this admission, Standard and Socony, who voluntarily signed the 'Red Line' agreement, opened themselves up to prosecution under the Sherman Act. Still unanswered is the question as to what, if any, promises of immunity were obtained from the US Government prior to using this argument as an excuse for abrogating the 'Red Line' agreement. The US Government, it should be noted, took no action against them at that time for signing it. In any event the two companies were able to break the contract without suffering any damage. CFP filed a suit against them. They filed a countersuit against CFP. Eventually the whole matter was settled out of court and the 'Red Line' agreement passed into history.[46]

There were several reasons why the United States allowed the agreement to be broken with impunity. One was the insistence of Ibn Saud that it be abrogated before Jersey and Socony buy into ARAMCO. The King's objections were outlined in a State Department memorandum, which noted:

> The King, however, is now much disturbed by the possibility that the Red Line Agreement may still be in force, or if not actually in force will still be honored in some way. What he fears is that this will prove an 'open door' through which British, French and other oil interests can get a foothold in the development of Arabian Oil. The Minister [Asad Al-Faquih, Saudi Arabian Minister to the United States] repeated twice that the King and the Crown Prince were agreed that the developments of Arabian oil resources must remain exclusively in American hands. He ended by saying that if the proposed new partners in Aramco could not disassociate themselves from previous commitments, they would not be permitted to buy into the Arabian concession.[47]

Despite the conflict between the King's demands and the traditional US insistence on the 'open door', the US was not at all averse to complying with Saud's demands. The reason was that the breaching of the agreement enabled the US companies to retain 100 per cent control of Saudi oil. On the other hand, if the 'Red Line' agreement had been observed, then the IPC would have joined the original partners in the Saudi concession. The US companies, it should be recalled, had only a 23.75 per cent interest in IPC. Thus the US ownership of Saudi crude would have been seriously diluted.

Foreign Aid. A second factor contributing to the displacement of Britain as the dominant oil power in the Middle East was the US foreign aid programme. As discussed earlier in this chapter, the US aid programme enabled many American firms to enter overseas markets. This was especially true of the Marshall Plan, whose chief beneficiary was the American oil industry and particularly those oil companies with holdings in the Middle East.

Between April 1948, when the Marshall Plan began operations, and November 1950 the US authorised the countries receiving aid to purchase $384 million of 'dollar' oil, i.e. oil produced by US companies in the Middle East.[48] It was this aid that opened up European markets to Saudi and other American-controlled Middle East oil. The importance of this assistance to both the US oil industry and the US Government cannot be overstressed.

Walter J. Levy, who resigned from Socony-Vacuum to head the oil division of the Marshall Plan, reported that without the Plan 'the American oil industry in Europe would already have been shot to pieces' and that the 'ECA [Economic Cooperation Administration, the official name of the Marshall Plan] has maintained outlets for American oil in Europe . . . which otherwise would have been lost'. He further stated that the 'ECA does not believe that Europe should save dollars by driving American oil from the European market'.[49] Levy, it should be noted, became the US oil industry's leading consultant after he left the ECA.[50]

In addition to paying for imported American-owned oil, the ECA helped the US oil industry by controlling refinery construction. Before World War II Europe had little refining capacity. Its imports consisted mainly of products. With the development of the post-war dollar shortage the European governments began a refinery construction programme to save foreign exchange. The savings from importing crude, rather than refined products, more than compensated for the foreign exchange costs of refinery construction.[51] The ECA, however, feared that these refineries might be used to import non-dollar crude and tried to limit their construction by restricting aid disbursements. Testifying before a Congressional committee in 1949, Dr Oscar Bransky, acting chief of the ECA's petroleum branch, described ECA oil policy as follows:

Under the stimulus of the shortage, the oil companies, British and American, both operating world-wide, had made plans in the postwar period to expand rapidly their crude-oil production and refinery capacity. The companies succeeded so well in carrying out their plans that by late 1948 supplies of many products were approximately in balance . . . There was no question that a large amount of expansion was justified in order to meet the expanding world demand, and there was also no question that it was ECA's assignment under a Congressional mandate to help the European countries save dollars where possible. Oil is the largest single item in the dollar budget of the participating countries, and savings of dollars on oil imports thus ranked high on the list of priorities of these countries.

But at the same time broad considerations of national policy made it essential that ECA do its best to prevent a degree of overexpansion that, because of the dollar shortage, would seriously reduce the market for American-owned oil and thus jeopardize American-owned concessions in foreign lands.[52]

Dr Bransky, who came to the ECA from Standard Oil of Indiana,[53] was presumably worried that the desire of European countries to reduce their imports of dollar oil would lower the demand for that oil and thereby endanger the US concession in Saudi Arabia. King Ibn Saud, it should be recalled, was then demanding that ARAMCO increase its production. Western Europe was the only logical market for this incremental production. If this market was pre-empted by the British, Dutch and French companies, ARAMCO could have lost its concession. To prevent this from happening the ECA was willing to increase the dollar cost of its programme even though it had a Congressional mandate 'to save dollars where possible'. In the words of Levy, where 'vital United States interests are at stake . . . the mere dollar-saving approach is too narrow in the case of oil'.[54] Levy, according to Bransky, refused to finance the refinery programme developed by the European countries.[55] Furthermore, he advised these countries that the 'ECA would be reluctant to sponsor export licenses for items of petroleum equipment'.[56] Such equipment was then in short supply. He finally induced the US Government to place a qualification on procurement authorisations of machinery under the aid programme, stating that 'such authorizations could not be used for petroleum equipment'.[57]

As a direct result of ECA's procurement and refinery policies, the Middle East displaced the Western Hemisphere as the chief source of Europe's oil. In 1947, before the start of the ECA, Europe received 43 per cent of its crude from the Middle East. In 1948, the first year of the Marshall Plan, the Middle East's share rose to 66 per cent. By 1950, it had grown to 85 per cent.[58] Europe had, to a significant degree, become dependent upon American-produced Middle Eastern oil. The full implications of this dependence did not become apparent until some 20 years later.

Before discussing these implications, we should examine some of the more immediate effects of this policy. It substantially enriched the oil companies with Middle East reserves. Their profits were so large that that they even enraged the US Ambassador to Egypt. In a 'restricted' message to the State Department, dated 16 November 1948, Ambassador Griffis took exception to a statement by the President of ARAMCO that the oil companies in the Middle East were financing a junior Marshall Plan there without any cost to the US taxpayer. His message stated:

> On the contrary, information at this end points to the vital conclusion that (1) since the US taxpayer supports the European Recovery program, (2) since ERP [European Recovery Program, another name for the Marshall Plan] oil purchases are 48% derived from U.S. oil companies' Middle East sources (may be 90% by 1951), and (3) since companies sell this oil at prices based on production costs in U.S.A. which are 100 to 500 per cent above Middle East production costs, it follows that via the exceptional profits derived from these excess prices, the U.S. taxpayer is contributing millions towards financing Middle East investments mentioned above.[59]

The point about the huge profits on Middle East oil was well taken. Under the basing point pricing system developed by the international oil companies, the price of Middle East oil was based on high-cost Texas oil. The oil companies charged this high price to the ECA while at the same time they charged their affiliates in the United States a lower price. In 1948 Middle East oil began to flow to the United States in substantial quantities. Given the freight rates and the New York market price at that time, the 'net back price' to the Middle East, i.e. the New York price less the US tariff less the tanker charges from the Middle East to New York, was $1.28 a barrel. The oil companies, at the same time, were charging the ECA $2.03 a barrel for Middle East oil shipped to Europe.[60] Throughout the entire period of the Marshall Plan there were similar discrepancies between the price that the oil companies charged the ECA and the price they charged other customers.

In 1952 the ECA filed a suit against the two original partners in ARAMCO, Standard Oil of California and the Texas Company, claiming overcharges of $100 million.[61] This suit, like so many similar actions against large, multinational corporations, dragged on in the courts for years. In 1956, four years later, the US Government amended its complaint and demanded $225 million in damages.[62] In

1957 Judge Thomas Murphy dismissed the government's complaint in a decision which, according to Engler, was based on 'a legalistic analysis that skirted most of the economic realities governing the industry'.[63]

Notwithstanding Engler's comments, it is questionable whether the government had a case, given the ambiguity of the ECA procurement regulations, many of which were drawn up when Levy was the oil administrator. Frank is undoubtedly quite correct in his observation that:

> The government's case was extremely weak . . . The great bulk of the shipments made by Socal and Texas Company to North America were clearly intra-company transactions and exchanges, which could hardly be considered open market sales. It may have been bad public relations to bill these at a figure below the price prevailing for Eastern Hemisphere sales, and perhaps also an offense against the spirit of ECA policy. But it hardly constituted a violation of the statute or the regulations, which were the only provisions legally binding on the participating suppliers.[64]

Questions can also be raised as to the zeal with which the government prosecuted its case. The original suit was filed when Truman was President. It continued under the Eisenhower regime, which had very close links to the oil companies. The actions of the government's attorneys left much to be desired. As Frank had commented, the 'ECA had little if any basis for a claim, and would most likely have lost the case, even it it had been handled brilliantly by the Government's attorneys, which was clearly not the case'.[65] One certainly has a right to wonder whether the whole prosecution was merely a charade which the government, for political reasons, felt compelled to act out but had no real intention of pursuing.

Wherever the truth may lie, Ambassador Griffis's comment that the US taxpayer, through the Marshall Plan, was 'contributing millions towards financing Middle East investments' was correct. These investments enabled the production of 'dollar' oil, i.e. oil extracted by US companies, in the Middle East to grow much more rapidly than non-dollar oil. Thus between 1947 and 1950 Middle East 'dollar oil' production rose by almost 150 per cent while that of non-dollar oil by only 85 per cent.[66] The Marshall Plan funds thus greatly accelerated the process of displacement of the latter oil.

Tax Assistance. The financial help extended to companies did not stop with the Marshall Plan and the other foreign aid programmes. The US Government used other forms of assistance as well. One was the decision of the US in 1944 to allow the Saudi Arabian Government to purchase gold from the US Treasury at a price of $35 dollars per ounce.[67] This decision greatly helped ARAMCO in a dispute it was having with the Saudis. Under the terms of the original concession agreement, ARAMCO was to pay Saudi Arabia a royalty of four gold shillings per ton of crude or its dollar or sterling equivalent.[68] The company paid royalties and advances to the government in gold sovereigns until the outbreak of World War II, when, because of embargoes, gold became unavailable. During the war it made payments in dollars. A controversy then arose over how gold should be valued in terms of dollars and pounds, i.e. over the dollar-gold, pound-gold and dollar-pound exchange rates. The company argued that the relevant rates were those in New York and London while the Saudis claimed that the rates in Jidda should apply. At that time the par value of the English gold pound in London was $8.2397. In Jidda the English gold pound was selling at between $16 and $20[69] or approximately twice the London price. Translated into dollar-gold equivalents, the London price was equal to $35 per ounce while the Jidda price fluctuated around $70 per ounce. By allowing the Saudis to use their dollars, which were paid to them by ARAMCO, to buy gold at $35 an ounce, the US relieved ARAMCO of paying the equivalent of $70 an ounce to the Saudi Treasury. It should be remembered that under the US laws existing at that time, the Treasury was forbidden to sell gold to individuals and corporations. It could sell gold only to 'friendly' governments. Saudi Arabia was obviously considered a 'friendly' government.

Another instance of financial help to ARAMCO came in 1949, when the Saudi Government was demanding increased payments from the company. The Saudi finance minister suggested that ARAMCO pay the Saudi Government an income tax and credit that tax against its US tax obligations.[70] Under the original concession agreement, ARAMCO was exempted from any income tax payments to Saudi Arabia. It was required to pay only a royalty. A royalty, technically, is not a tax but a payment for the right to extract the oil. Under the US income tax laws, royalties were considered an expense and, as such, could not be fully written-off against US tax obligations. If, for instance, the corporate income tax rate were 50 per cent and the increase in royalties were $1 million, then the company could reduce its US tax obligations by only $500,000, i.e. by 50 per cent of $1 million. An increase in income taxes of $1 million could, on the other hand, be fully written-off against US tax obligations.

The problem for ARAMCO was that, while the US Government allowed its companies to write-off their foreign income tax payments, these write-offs had to be genuine. Since ARAMCO was exempt from any income tax payments, the Saudi Government could not presumably impose an income tax on ARAMCO without its consent. If ARAMCO gave its consent, ARAMCO officials feared that the US tax authorities might view this consent as a means to evade US taxes. The authorities could, after all, point out that ARAMCO could increase its royalties to the Saudis to satisfy their demand for more income. In addition ARAMCO was concerned that the US tax authorities might view with suspicion the fact that ARAMCO financial experts had examined Saudi Arabia's finances and were in a position to shape that country's tax structure.[71]

In July 1949 the officials of ARAMCO voiced their concerns to the State Department. They met with a group headed by George McGhee, Assistant Secretary of State for Near East, South Asian and African Affairs. McGhee was himself an oil man. He was sole owner of McGhee Production Company, an independent oil company.[72] He had been previously Treasurer of the Creole Oil Company in Venezuela,[73] a Rockefeller-controlled company whose parent, Jersey Standard, was a partner in ARAMCO. Upon leaving the State Department he became a director of Mobil,[74] also a partner in ARAMCO. McGhee was thus in a position to understand fully the problems faced by ARAMCO.

While not committing himself, he promised that 'the Department would move as rapidly as possible in the matter'.[75] Two days later, a Department group met, discussed the matter and forwarded to McGhee a recommendation 'that ARAMCO should be advised it was the Department's view that the problem was one which primarily concerned ARAMCO, SAG [Saudi Arabian Government], and the U.S. Treasury'. The group further suggested the Department informally tell ARAMCO that 'if SAG should ask for guidance in tax matters the Department of State would pursue its traditional policy of assisting foreign governments in obtaining competent and independent advisers on tax matters'.[76] Thus, the substitution of 'competent and independent advisers' for ARAMCO's experts presumably would clear ARAMCO of suspicion that it was structuring the Saudi tax laws in such a way as to avoid the payment of US income taxes. The Department group made this recommendation, even though one of its members, Gordon

Mattison, pointed out that compliance with ARAMCO's request 'might cost the U.S. Treasury an estimated $60,000,000 a year'.[77]

After receiving this recommendation, McGhee despatched a letter to John Graham, Assistant Secretary of the Treasury, explaining ARAMCO's problem. McGhee wrote:

> The essence of the matter is that the oil company has a clause in its concession exempting it from paying income tax to the Saudi Arabian Government, and the Arabian-American Oil Company must now decide whether or not to waive this clause. Aramco's primary interest is whether if such a waiver were granted it would be able to deduct income tax paid to the Saudi Arabian Government from income tax due to the United States Government. There is the added problem of whether or not the Saudi Arabian Government should be given expert advice in drafting a proposed income tax law.[78]

It is interesting that McGhee wrote this letter, which followed a phone call to Graham, despite the recommendation of the State Department group that 'the problem was one which primarily concerned ARAMCO, SAG, and the US Treasury'. McGhee's letter and phone call indicated that it concerned the State Department as well.

Soon afterwards the Treasury and ARAMCO sent a team of tax lawyers to Saudi Arabia to draw up a corporate income tax law in a country which had very few corporations and no income tax. They reached an agreement in 1950, under which ARAMCO agreed to submit to a Saudi income tax.[79] In 1951 the partners in ARAMCO began claiming substantial US tax credits for their payments to the Saudi Government. Apparently there was a dispute in the Treasury Department over the legitimacy of these tax credit claims. Finally, in 1953 or 1954 the Treasury, in a *private* ruling, upheld the claims. This ruling was not published until 1955.[80]

The Treasury's actions were most unusual. They caused a substantial loss to the US Treasury. Dr John Blair, who served as Chief Economist of the US Senate Subcommittee on Antitrust and Monopoly and was co-author of the US Federal Trade Commission's classic study, *The International Petroleum Cartel*, commented:

> Between 1950 and 1951, Aramco's payment to Saudi Arabia rose from $66 million to $110 million, while its payment to the US Treasury fell from $50 million to $6 million. The only loser was the

American taxpayer, who had to make up the loss in tax revenues. By 1955, the annual loss to the US Treasury from Aramco had tripled, rising to $154 million. Comparing the company's actual income statement with what would have been the case had it been a domestic U.S. producer, assuming the other entries to have been the same, Aramco, as a domestic concern, would have returned $154,867,000 in taxes to the federal government plus $42,030,000 to state and local governments. Instead $192,743,000 was paid in the form of corporate income taxes to the Saudi Arabian Government.[81]

These tax concessions enabled ARAMCO to satisfy the Saudi Government's demand for higher revenue, actually 50 per cent of ARAMCO's profits, without costing the company one penny. At the time these concessions were granted the company was not exactly impoverished. According to Blair, before the imposition of the Saudi income tax the company retained 91¢ a barrel, after paying operating expenses of 20¢, a royalty to the Saudi Government of 21¢ and a US income tax of 43¢.[82] In 1950, the last year under this system, ARAMCO produced 200 million barrels of oil,[83] thus receiving, after paying all operating costs, royalties and taxes, a cash flow[84] of $182 million. Jersey Standard and Socony-Vacuum, it should be recalled, paid out $125 million to buy into ARAMCO, which sum enabled the original partners to recoup their investment. An annual cash flow of $182 million on an investment of $125 million does not imply a low rate of return. Had ARAMCO acceded to the Saudi's Government's demand through an increase in royalties, rather than through the imposition of an income tax, it would have retained 50¢ a barrel after paying operating costs, a royalty to the Saudis of 88¢ and a U.S. income tax of 17¢.[85] On the same production it would have kept $100 million. A cash flow of $100 million on an investment of $125 million still would not have left the company poverty-stricken. As it turned out, after the imposition of the Saudi income tax, ARAMCO retained 91¢ per barrel after paying 21¢ in royalties to the Saudi Government, 67¢ in income taxes to that government and receiving from the US Government an 'excess tax credit' of 24¢, which it used to reduce its US tax obligations on income earned in other parts of the world.[86] Its total cash flow on its 1950 production was exactly the same as under the original pure royalty system.

What was unusual about the Treasury's actions was that they were done in private without the consent of Congress. What, in effect, was a significant appropriation of public funds was performed by a few individuals even though, under the US Constitution, all such appropriations are supposed to be approved by Congress. Surprisingly, this action was never seriously questioned until some 20 years later when Senator Frank Church, in his capacity as Chairman of the Subcommittee on Multinational Corporations, asked George McGhee:

> And I am wondering if the Congress was ever consulted. This would seem to me to be a very ingenious way of transferring many millions by executive decision out of the public treasury and into the hands of a foreign government treasury without ever needing any appropriation or authorization from the Congress of the United States. Isn't that true? Or was Congress asked to authorize this transfer.[87]

All McGhee could say was that 'there was, to the best of my knowledge, consultation between the Treasury Department and the appropriate congressional committee'.[88] What was even more remarkable were McGhee's apparent lapses of memory. When, for instance, Jack Blum, Associate Committee Counsel, asked him if ARAMCO officials suggested that 'if the U.S. Government wanted more money paid the U.S. Government should pick up some of that tab . . . was that part of the conversation', McGhee replied that 'they didn't go into any detail on a tax advantage with us'.[89] Several years after McGhee made this statement in 1974, the State Department declassified some hitherto secret material. Below are excerpts from the Department's summary of the meeting between McGhee, other State Department personnel and ARAMCO officials on 20 July 1949, to which we already referred:

> Mr. Duce [Vice-President of ARAMCO] said that about a year and a half ago Prince Faisal when he was in America had remarked that Aramco paid a large income tax to the United States Government and should pay some tax to the Saudi Arabian Government. Speaking in the same vein the Saudi Arabian Finance Minister told Mr. Duce last January that the Government of Saudi Arabia needed money. *He said he understood that Aramco could pay Saudi Arabia an income tax and deduct it from the company's U.S. income tax, thus putting no increased burden on Aramco but shifting the flow of dollars from the U.S. Treasury to the Treasury of the Saudi Arabian Government . . .* [Italics added]

[Mr. Brougham, Vice-President, Financial Matters of ARAMCO] went on to say that the Saudis had asked George Eddy, of the Treasury Department, when he was in Saudi Arabia how to get more money out of oil, and Eddy mentioned various methods such as increasing royalties, participating in profits, and the possibility of Saudi Arabia levying an income tax on Aramco. *It seems that Eddy told the Saudis that the sums paid by Aramco to Saudi Arabia in income tax would be deductable* [sic] *from U.S. income taxes paid by the company . . .* [Italics added]

When Mr. McGhee suggested that the question be put up to the Bureau of Internal Revenue Mr Erskine [an attorney for ARAMCO] said the Bureau would probably not give a ruling unless the case was very specific and the terms of the new tax had been worked out in detail. [Italics added]

Mr. Brougham said that the company would like to advise and work with the Saudi Arabian Government through recognized experts as has been done earlier this year when Aramco through Judge Hudson guided Saudi Arabia in formulating a policy for offshore oil. Aramco was afraid, however, that if they did this the Department of Justice would say that the tax was not a genuine one and would disallow Aramco obtaining an exemption on it.[90]

When asked by a Committee staff member about how the State Department conveyed its views on the matter to the Treasury, McGhee replied:

The department's foreign policy objectives would have been conveyed through the National Security Council. I assume that the Treasury Department had representatives at whatever large meetings we had at my level in the department. *I don't recall myself ever going to the Treasury Department.* [Italics added][91]

While there is no evidence that McGhee ever physically went over to the Treasury Department to discuss this problem, he, as has already been noted, both telephoned and wrote to the Assistant Secretary of the Treasury about ARAMCO's dilemma. For some unexplained reason he failed to recall his telephone conversation and letter even though he looked through his records before testifying.[92] Whatever may have been Mr McGhee's reasons in not divulging all the information, he justified the tax decision on the grounds that it was 'exceedingly important from the standpoint of the stability of regimes in the area and the security of the Middle East as a whole and the continued ownership of our oil concessions there and the ability to exploit them that the Government of Saudi Arabia receive an increased oil income'.[93] Apparently the interests of the United States coincided with those of the feudal Saudi regime and the profits of ARAMCO.

Needless to say, ARAMCO was not the only oil company to enjoy the largesse of the US Treasury. Royalties were transformed first into income taxes and subsequently into foreign tax credits in Kuwait, Iran and other places.[94] There were no apparent limits to the US Government's generosity to the oil companies, the flag-bearers of the new American Empire.

In addition to the aid given the companies through the Marshall Plan, the sale of gold and the foreign income tax credits, the US helped them through the exertion of diplomatic pressure against Britain. In 1950 Britain, in an effort to reduce her dollar deficit, took measures to favour the importation of 'sterling' oil, i.e. oil produced by Anglo-Iranian and Shell, over 'dollar' oil.[95] The State Department, expressing its 'emphatic concern', immediately protested.[96] After several months of negotiations, the United States and the United Kingdom reached a compromise under which Britain agreed to ease its restrictions and the US oil companies consented to accept partial payment in sterling.[97]

Iran. The US Government also played a direct role in opening up Iran to American oil companies, which, until 1951, had been the private preserve of Anglo-Iranian. The role of the United States in overthrowing Mossadegh and returning the Shah has been well documented and has received relatively widespread publicity. The US moves in Iran have generally been interpreted by defenders and critics alike as part of the US Cold War strategy. The general presumption is that the US feared Mossadegh would ally himself with the Soviet Union and therefore moved to overthrow him and replace him with the more reliable Shah.[98] While this is unquestionably true, it is only part of the story. What is generally not so well recognised is that the US action was directed also against Britain. It was the means by which the US was able to break the British oil monopoly in that country.

When Mossadegh nationalised the properties of Anglo-Iranian in 1951, the United States adopted an ambiguous attitude towards this move. On the one hand, as the nation with the largest stake in

overseas investment, it expressed concern about the expropriation.
John Loftus, head of the State Department's petroleum division,
complained:

> While recognizing the sovereign right of any country to assume
> ownership . . . of the petroleum industry or any of its branches,
> this Government must nevertheless recognize and proclaim that
> international commerce, predicated upon free trade and private
> enterprise (which is the conceptual core of United States economic
> foreign policy), is, in the long run, incompatible with an extensive
> spread of state ownership and operation of commercial properties.[99]

On the other hand, the State Department had no intention of using its
influence to restore the seized properties to Anglo-Iranian. It vetoed
an attempt by the British to take military action against Iran.[100]
Unquestionably the Department feared that such a move might
provoke Soviet intervention. But also it did not want the British to
regain their former position in the country. As Engler has noted, the
nationalisation presented the US with 'a new opportunity'[101] to bring
the 'open-door' policy to Iran.

The overthrow of Mossadegh, which was engineered by the CIA,
paved the way for the displacement of Britain by the United States
as the major power in Iran. The displacement took place first in oil
and then in the other sectors of the economy.

After the coup the US sent Herbert Hoover, Jr, a director of Union
Oil,[102] to negotiate a new oil pact. Under the agreement the properties
formerly belonging to Anglo-Iranian were turned over to the NIOC
(National Iranian Oil Company), a company wholly owned by the
Iranian government. Oil exploration and production in southern Iran
and the operation of the Abadan refinery, then the world's largest,
were however to be carried out by a consortium of companies[103] known
as Iranian Oil Participants Ltd under contract to the NIOC.[104] As
originally envisaged the membership in the consortium was to be
restricted to the eight major international oil companies, the cartel
companies,[105] namely, Anglo-Iranian, whose name had by then been
changed to BP (British Petroleum), Gulf, Socony, Standard Oil (New
Jersey), Standard Oil of California, Shell, Texaco and CFP. BP was to
receive 40 per cent, Shell, 14 per cent, CFP, 6 per cent and the other
five companies, all American, were to receive 8 per cent each. Thus, on
a national basis the shares of the consortium were distributed as follows:
England, 46 per cent, United States 40 per cent, Holland 8 per cent

and France, 6 per cent.[106]

There was, however, one important problem with the share going to the five US companies. All these companies at that time were undergoing prosecution for violation of the US anti-trust laws in what has been referred to as the 'Cartel Case'.[107] This case, initiated by the Department of Justice in 1952, was partially based on the findings of the Federal Trade Commission's famous report, *The International Petroleum Cartel*. Among the stated objectives of the Department was the ending of the 'monopolistic control' of foreign production and of the policy of 'excluding American independents from foreign sources of supply'.[108] Thus, at the very moment at which the Justice Department was pressing this suit, the State Department negotiated a treaty which gave the American cartel companies a greater control of foreign production than ever before and which excluded American independent companies from Iran. The agreement with Iran was, in fact, signed by Harold W. Page, Vice-President, Standard Oil (New Jersey), who represented the consortium in the negotiations.[109]

This incongruity, however, did not deter US policy-makers. The Justice Department merely issued an advisory opinion that the agreement did not violate the anti-trust laws.[110] To make their highly imaginative interpretation of the law a little more palatable, the Department's officials persuaded each of the five US companies to surrender a one per cent share to American independent companies. Thus, the independents, as a group, were to receive a 5 per cent share and each of the five US internationals was to receive 7 per cent.[111]

Once this decision was made the question arose on how to select the independent companies who were to participate in the consortium. Since the State Department did not want to be put in the position of favouring one US firm over another, it selected the accounting firm of Price Waterhouse and Company to determine the eligibility of the applicant firms. That company undoubtedly had the qualifications to perform the screening process. Among its clients were Standard Oil (New Jersey), Standard of California, Gulf and Shell.[112] The choice of this company evoked a bitter protest from Watson Snyder, the petroleum specialist in the Justice Department's Anti-trust Division. Snyder complained:

all through the documentary material delivered by the five defendants in the cartel case, you will find that Price Waterhouse and Company is the medium through which all the accounting is done for the participants in the various illegal arrangements . . .

Whenever either the domestic or foreign branches of the petroleum industry carry out any joint operations Price Waterhouse is chosen to do the accounting'.[113]

As Blair pointed out, 'Price Waterhouse was not likely to certify any firm strongly objected to by the majors'.[114] This incident illustrates both the intra-industry problems generated by the policy of supporting private American firms in their quest for overseas reserves and the close links between the State Department and the large international oil companies.

Price Waterhouse performed its duty faithfully. It approved every US applicant except one, the International Cooperative Petroleum Association. It vetoed this applicant on the grounds that the Association was not truly an American company 'in that it has members in other countries'. There is however evidence that the international oil companies, which were consulted by Price Waterhouse, feared that the Association's 'participation in the sale of Iranian oil in Western Europe could upset prevailing price structures'.[115] Originally Price Waterhouse approved twelve US firms, who together formed Iricon Agency Limited, which held 5 per cent of the shares in the consortium. Mergers and sales of shares eventually reduced the number of participants in Iricon to six. In 1974 they were Atlantic Richfield, which owned one-third of Iricon's shares; American Independent, Getty and Charter, each of which owned one-sixth; and Sohio and Conoco, each of which owned one-twelfth.[116]

The exemption of the consortium members from anti-trust action spelled the doom of the 'Cartel Case' against the five US internationals. The case originally covered all phases of the industry's entire international operations. But the charges relating to monopolistic practices in production, transportation, storage and refining were dropped on the grounds that the prevailing practices were less restrictive than those countenanced in the consortium agreement. Hence the exemption of that agreement implied the exemption of the other practices. The Antitrust Division dismissed all charges without prejudice against Mobil and Standard Oil of California. The whole case ended in 1968 without any substantial impact on the industry.[117]

There were several important repercussions following the establishment of the consortium. First, it set up for the first time a joint venture in which all the eight cartel companies participated. Because of this particular characteristic of the consortium, Iran played the pivotal role in determining world production. Once Iranian production was set, all the partners in the consortium set their production in other parts of the world accordingly. The companies used a formula, APQ (Average Program Quantity), through which they coordinated Iranian production with production in the other countries in which they had concessions.[118] This was a cartel *par excellence*.

Second, though BP lost its monopoly in Iran, it did not lose financially. The consortium members estimated the value of BP's assets at $1 billion. Accordingly the other participants paid BP $600 million for their 60 per cent share in the consortium. The Iranian Government, under the newly-restored Shah, also agreed to pay BP $70 million as indemnity.[119] This indemnity was a major factor in inflaming Iranian opinion against the consortium and against the Shah. There was popular feeling that BP had taken far more than its just share of profits long before the nationalisation. There was no justification to pay it any indemnity. By accepting this indemnity, BP, along with the other members of the consortium, help lay the seeds of the Islamic Revolution which was to sweep the country some 25 years later.

Third, the establishment of the consortium was the most important factor in making the US the dominant oil power in the Middle East. The entry into Iran unquestionably gave it effective control of most of the known reserves in the non-Communist world. Its basic objective, the control of world oil, had been realised. This was to have important implications, not only in the Middle East, but throughout the world.

Notes

1. N. Jacoby, *Multinational Oil, A Study in Industrial Dynamics* (Macmillan, New York, 1974), p.37.
2. US Senate, Subcommittee on Multinational Corporations of the Committee on Foreign Relations, *A Documentary History of the Petroleum Reserves Corporation, 1943-1944*, 93rd Congress, 2nd Session (US Government Printing Office, Washington, 1974), p.3.
3. *Historical Statistics of the United States, Colonial Times to 1957: A Statistical Abstract Supplement* (US Government Printing Office, Washington, 1960), p.143.
4. Ibid., p.735.
5. Ibid.
6. Among the works dealing with this aspect of the origins of the Cold War are G. Alperovitz, *Atomic Diplomacy: Hiroshima and Potsdam, the Use of the Atomic Bomb and the American Confrontation with Soviet Power* (Simon

& Schuster, New York, 1965); D. Fleming, *The Cold War and Its Origins 1917-1960* (Allen & Unwin, London, 1961); D. Horowitz, *The Free World Colossus: A Critique of American Foreign Policy in the Cold War* (Hill & Wang, New York, 1965); G. Kolko, *The Politics of War: the World and United States Foreign Policy 1943-5* (Vintage Books, New York, 1968); G. Kolko, *The Roots of American Foreign Policy: An Analysis of Power and Purpose* (Beacon Press, Boston, 1969); R. Maddox, *The New Left and the Origins of the Cold War* (Princeton University Press, Princeton, N.J., 1973); J. Morray, *From Yalta to Disarmament, Cold War Debate* (MR Press, New York, 1961); W. A. Williams, *The Tragedy of American Diplomacy* (Dell Press, New York, 1962).

7. H. Magdoff, *The Age of Imperialism, The Economics of U.S. Foreign Policy* (Modern Reader Paperbacks, New York and London, 1969), p.42.

8. P. B. Kenen, *Giant Among Nations, Problems in United States Foreign Economic Policy* (Rand McNally, Chicago, 1960), p.159.

9. H. Magdoff, *The Age of Imperialism*, p.126.

10. US Senate, Committee on Foreign Relations, *Hearings on Foreign Assistance of 1962* (US Government Printing Office, Washington, 1962), p.27.

11. U.S. Senate, Committee on Foreign Relations, *Some Important Issues in Foreign Aid* (US Government Printing Office, Washington, 1966), p.15.

12. The trade balance is the difference between *merchandise* exports and *merchandise* imports. As such, it should be distinguished from other balances, such as the balance of payments, which encompasses all international transactions including service and capital flows as well as merchandise movements.

13. The balance-of-goods-and-services encompasses the trade balance and and the balance on services, like insurance, travel, freight and investment income.

14. *Economic Report of the President Transmitted to the Congress February 1974* (US Government Printing Office, Washington, 1974), p.352.

15. *Economic Report of the President Transmitted to the Congress January 1980* (US Government Printing Office, Washington, 1980), p.318.

16. Ibid.

17. M. Wilkins, *The Maturing of Multinational Enterprise: American Business Abroad from 1914 to 1970* (Harvard University Press, Cambridge, Mass., 1974), p.329.

18. US Senate, *A Documentary History of the Petroleum Reserves Corporation*, p.3.

19. Ibid.

20. Ibid., p.4.

21. Ibid.

22. Ibid., p.5.

23. Ibid.

24. Ibid., p.6.

25. Ibid., p.16.

26. Ibid., p.35.

27. Ibid., p.v.

28. Ibid., p.61.

29. Ibid., p.63.

30. Ibid.

31. Ibid., p.65.

32. Ibid., p.63.

33. Ibid.

34. W. J. Kennedy (ed.), *Secret History of the Oil Companies in the Middle East* (2 vols., Documentary Publications, Salisbury, N.C., 1979), vol. 1, p.104.

35. Ibid., p.105.

36. Ibid., p.104.

37. Ibid., p.8.
38. Ibid., p.17.
39. Jacoby, *Multinational Oil*, p.39.
40. US Federal Trade Commission, *The International Petroleum Cartel* (US Government Printing Office, Washington, 1952), p.120.
41. C. Issawi and M. Yeganeh, *The Economics of Middle Eastern Oil* (Praeger, New York, 1962), p.180.
42. Kennedy, *Secret History of the Oil Companies in the Middle East*, vol. 1, pp.75-6.
43. US Federal Trade Commission, *The International Petroleum Cartel*, p.103.
44. Ibid., p.104.
45. Ibid.
46. Ibid., pp.103-5.
47. Kennedy, *Secret History of the Oil Companies in the Middle East*, vol. 1, p.77.
48. E. Groen, 'The Significance of the Marshall Plan for the Petroleum Industry in Europe – Historical Review of the Period 1947-1950' in US Congress, *The Third World Petroleum Congress* (US Government Printing Office, Washington, 1952), p.49.
49. J. Kolko and G. Kolko, *The Limits of Power, The World and United States Foreign Policy, 1945-1954* (Harper and Row, New York, 1972), p.447.
50. J. Stork, *Middle East Oil and the Energy Crisis* (Monthly Review Press, New York and London, 1975), p.61.
51. P. Odell, *Oil and World Power: Background to the Oil Crisis*, 3rd edn (Penguin Books, Harmondsworth, 1974), p.99.
52. US House of Representatives, Select Committee on Small Business, *Hearings, Effects of Foreign Oil Imports on Domestic Producers*, 81st Congress, 1st Session (US Government Printing Office, Washington, 1949), part 2, p.523.
53. Ibid., p.522.
54. Ibid., p.531.
55. The ECA did eventually finance some expansion in refining capacity, including capacity by foreign companies. It did so because, in the words of Bransky, any cut-off of funds would have probably resulted in 'a drastic limitation of dollar oil imports resulting in an even greater increase in expansion of non-dollar projects'. Though the European countries wanted to increase capacity from 480,000 barrels per day to well over one million barrels, the ECA refused to fund any expansion above 840,000 barrels. In addition the ECA refused to finance any government-owned refineries. See ibid., p.532.
56. Ibid., p.530.
57. Ibid., pp.530-1.
58. E. Shaffer, *The Oil Import Program of the United States, An Evaluation* (Praeger, New York, 1968), p.13.
59. Kennedy, *Secret History of the Oil Companies in the Middle East*, vol. 2, p.245.
60. H. Frank, 'The Pricing of Middle East Crude Oil', PhD thesis, Columbia University, 1961, p.58.
61. R. Engler, *The Politics of Oil, A Study of Private Power and Democratic Directions* (The Macmillan Company, New York, 1961), p.220.
62. Ibid., pp.522-3.
63. Ibid., p.523.
64. Frank, 'The Pricing of Middle East Crude Oil', p.86.
65. Ibid.
66. E. Shaffer, 'The Oil Import Program of the United States: An Evaluation', PhD thesis, Columbia University, 1966, p.65.
67. Kennedy, *Secret History of the Oil Companies in the Middle East*, vol. 1, p.86.

68. Ibid., pp.81-2.

69. Ibid., p.81.

70. Ibid., vol. 2, p.425.

71. Ibid., pp.425-6.

72. US Senate, Subcommittee on Multinational Corporations of the Committee on Foreign Relations, Hearings, *Multinational Corporations and United States Foreign Policy*, 93rd Congress, 2nd Session (US Government Printing Office, Washington, 1974), part 4, p.83.

73. Kennedy, *Secret History of the Oil Companies in the Middle East*, vol. 2, p.428.

74. L. Turner, *Oil Companies in the International System* (The Royal Institute of International Affairs, London, 1978), p.104.

75. Kennedy, *Secret History of the Oil Companies in the Middle East*, vol. 2, p.429.

76. Ibid., p.432.

77. Ibid., p.430.

78. Ibid., p.433.

79. J. Blair, *The Control of Oil* (Pantheon, New York, 1976), p.199.

80. US Senate, *Multinational Corporations and United States Foreign Policy*, part 4, p.124.

81. Blair, *The Control of Oil*, pp.200, 202.

82. Ibid., p.201.

83. Issawi and Yeganeh, *The Economics of Middle East Oil*, p.183.

84. The cash flow of a company is equal to after-tax profits plus capital consumption allowances. The cash flow is what is left, after paying all current expenses and taxes, for either distribution to shareholders as dividends or investment. In their accounting oil companies used the 'discounted cash flow', rather than profits *per se*, to assess the viability of any given investment. See M. Tanzer, *The Political Economy of International Oil and the Underdeveloped Countries* (Beacon Press, Boston, 1969), pp.32-3.

85. Blair, *The Control of Oil*, p.201.

86. Ibid., pp.200-1.

87. US Senate, *Multinational Corporations and United States Foreign Policy*, part 4, p.89.

88. Ibid.

89. Ibid., p.92.

90. Kennedy, *Secret History of the Oil Companies in the Middle East*, vol. 2, pp.425-7.

91. US Senate, *Multinational Corporations and United States Foreign Policy*, part 4, p.90.

92. Ibid., p.88.

93. Ibid., p.89.

94. Blair, *The Control of Oil*, p.199.

95. H. Menderhausen, *Dollar Shortage and Oil Surplus in 1949-1950* (Princeton University Press, Princeton, 1950), p.2.

96. Ibid., p.11.

97. Ibid., pp.28-32.

98. See, for instance, F. Halliday, *Iran Dictatorship and Development* (Penguin Books, Harmondsworth, 1979), p.25 and Stork, *Middle East Oil and the Energy Crisis*, p.52.

99. Stork, *Middle East Oil and the Energy Crisis*, pp.51-2.

100. Ibid., p.50.

101. Engler, *The Politics of Oil*, pp.202-3.

102. Turner, *Oil Companies in the International System*, p.104.

103. Jacoby, *Multinational Oil*, p.97.

104. Issawi and Yeganeh, *The Economics of Middle East Oil*, p.175.

105. See US Federal Trade Commission, *The International Petroleum Cartel.*

106. J. Walden, 'The International Petroleum Cartel in Iran — Private Power and the Public Interest', *Journal of Public Law*, vol.2, no. 1 (1962), p.49.

107. United States v Standard Oil Company (NJ) 86 Civ. 27, (SDNY).

108. Blair, *The Control of Oil*, p.72.

109. Engler, *The Politics of Oil*, p.208.

110. Walden, 'The International Petroleum Cartel in Iran — Private Power and the Public Interest', p.48.

111. Ibid., p.150.

112. H. O'Connor, *The Empire of Oil* (Monthly Review Press, New York, 1962), p.330.

113. Blair, *The Control of Oil*, p.45.

114. Ibid.

115. Walden, 'The International Petroleum Cartel in Iran — Private Power and Public Interest', p.50.

116. Blair, *The Control of Oil*, p.46.

117. US Senate, *Multinational Corporations and United States Foreign Policy*, part 4, p.189.

118. Blair, *The Control of Oil*, pp.101-8.

119. Walden, 'The International Petroleum Cartel in Iran — Private Power and Public Interest', p.52.

6 THE CONTROL OF WORLD OIL: POLICY AND PROBLEMS

That the very success of US oil policy in Iran carried the seeds of its own destruction was not apparent to US policy-makers until some two decades later. More immediately apparent were the interrelated problems of State Department relations with individual oil companies and the management of the oil surplus that was then engulfing the world.

Relationship with Individual Companies

Iran illustrated the dilemma the US Government faced in implementing its policy of encouraging US nationals to find reserves overseas. In a country with many oil companies such a policy must inevitably favour some companies over others. Logically the national interest would lie in having a few firms engage in foreign oil operations. It would then be relatively easy for the State Department to monitor their activities and to coordinate them with the Department's policies. The entry of a large number of firms into overseas fields would render such monitoring and coordination difficult.

The logic of foreign policy however conflicted with the dogma of American free enterprise. This dogma, as enshrined in the anti-trust laws, dictated that the government should play no favourites among companies, except, perhaps, to aid small firms in their competitive battles with large ones.[1] The exigencies of foreign policy, on the other hand, demanded that government should deal chiefly with those enterprises experienced in foreign operations, i.e. with the large international oil companies.

This conflict was exemplified by the original attempt of the State Department to invite only the five large US internationals to participate in the Iranian consortium. Pressure both from other oil companies and the Antitrust Division of the US Department of Justice forced the State Department to allow some smaller American companies to buy into the consortium. The State Department, as has been pointed out, appointed the accounting firm of Price Waterhouse and Company to screen the applicants in order to preserve its official

113

position of maintaining neutrality among competing American firms. In
reality this appointment of a company with close links to the
internationals indicated the Department's bias towards these firms.

Control of Entry

Because diplomatic protection was an essential ingredient of an
investment decision in many countries, the Department was in a position
to control entry into foreign oil. Firms entering an area without the
Department's blessing did so at their own risk. Generally the Department
used its power to protect the internationals against potential competitors.
This screening process was most evident in Saudi Arabia and in other
areas of the Middle East. An example can be seen in the treatment of
Superior Oil.

Saudi Arabia. In July 1947 three executives of Superior Oil arrived in
Jidda, Saudi Arabia seeking to meet government officials to discuss
possible concession terms. Though they had telegraphed in advance to
the US Minister, J. Rives Childs, the time of their arrival, the Minister
chose not to meet them at the airport, sending instead his third
secretary. When William Keck, the Executive Officer of Superior, asked
Childs to arrange an appointment to meet King Ibn Saud, Childs
refused to do so. In a despatch to the State Department Childs wrote:
'I informed Keck that it was unnecessary; that he need have no fear
but that the King was aware of the presence of the Superior Oil
Company group in Saudi Arabia'.[2] Despite Childs' misgivings about
the Superior contingent,[3] he arranged for them to meet several local
Saudi officials. He also entertained them for dinner, inviting, among
others, the local representative of ARAMCO. In his report to the State
Department Childs explained:

> I thought it important to include Mr. Campbell of Aramco, in
> order not to give the impression to the Saudi Arabian
> Government that we were playing up the Superior Oil Company
> in the midst of Aramco's difficulties in a way which might confuse
> the Saudi Arabian authorities. I believe Mr. Campbell appreciated
> the care I took in this regard.[4]

The problems to which Childs referred related to the location of
ARAMCO's proposed pipeline to the Mediterranean, the financing
and operations of railroads and piers, the gold-pound controversy
and the Saudi demand for increased royalty payments. Of particular

concern to both ARAMCO and the US Government at that time was the location of the terminus of the pipeline. The construction of the line was considered a top priority to provide outlets both to satisfy King Ibn Saud's demands for increased production and to replace Western Hemisphere oil in European markets with oil from the Middle East. Its construction was being delayed because of real estate speculation by Saudi officials. A 'secret' State Department memo of 6 June 1947, for instance, revealed:

> The exact location of the terminus of the Trans-Arabian pipeline on the eastern shore of the Mediterranean has been complicated by the fact that various Saudi Arabian officials have been speculating in land at the places where they thought the line would end.[5]

Needless to say, the necessity of mollifying the conflicting Saudi interests involved ARAMCO in delicate negotiations with various layers of Saudi officialdom. The entrance of an outsider, who might offer these officials a different set of proposals, could obviously upset these negotiations. Such a development would cause ARAMCO justified concern. Whether it should also cause concern to the US Government is, of course, a matter of contention.

The officials of Superior, quite understandably, did not sympathise with the concern expressed by the US Minister. In a meeting with State Department officials on 6 May 1948, the company's Vice-President, Carlton Wood, voiced dissatisfaction with Childs's treatment of Superior's personnel. He complained specifically that Childs would not grant the company's representatives an interview, refused to house them in the Legation and discussed Superior's plans with ARAMCO officials. Loy Henderson, the Department's Director of Near Eastern and African Affairs, assured Wood that 'Mr. Childs intended no such discrimination but was probably quite apprehensive lest any action on his part be construed as indicating official US Government sympathy to one party or the other.'[6]

What Henderson neglected to mention was that the Department apparently was not concerned about giving the appearance of favouritism when it dealt with ARAMCO. When Childs, for instance, notified the Department on 20 March 1947, just two months before the arrival of the Superior group, that 'Mr. D. C. Dennett, of the American Legation in Beirut, arrived in Jidda on March 14 1947, *in an Aramco plane from Dhahran on a brief tour of the Middle*

East during which he travelled with Mr. William Lenahan of the Arabian American Oil Company '[7] [italics added] , he expressed no misgivings about the impression such travel arrangements might have on the Saudi Government. Nor apparently did the State Department.

The Department still continued to profess its neutrality. This professed neutrality was however frequently questioned by Superior. In a meeting with Department officials on 17 June 1948, Robert Allen, Jr, Superior's Vice-President in charge of foreign operations, complained that ARAMCO officials knew about his company's discussions with the Saudi Government. He wanted to know whether the Department was really neutral in its treatment of ARAMCO and Superior.

According to the Department's memorandum of the meeting, Henderson told him that:

> the Department continued to favor competition and the appearance of new companies as concession holders in the Middle East but could not to this end favor any one company over another. The Department has endeavored to treat all companies alike and felt that this position must be maintained.[8]

This type of neutrality, to the extent that it existed, was, of course, no neutrality at all. It favoured the powerful, established firms over the newcomers. It is reminiscent of Anatole France's famous dictum, 'the law in its majestic equality forbids the rich as well as the poor to sleep under the bridges, to beg in the streets and to steal bread'.[9] But, despite its professions to the contrary, the State Department didn't even practise this type of neutrality. It took positive action on behalf of ARAMCO.

ARAMCO and Superior were involved in a dispute over the right to develop Saudi Arabia's off-shore oil resources. Superior had been seeking off-shore concession rights from the Saudi Government. The State Department was opposed to the granting of these rights until an agreement could be reached with Britain and the states bordering the Persian Gulf on how to apportion the sub-surface rights among the interested parties. The ceding of these rights by Saudi Arabia to Superior or any other oil company would jeopardise the negotiations leading to such an agreement. In addition, there was the question as to whether ARAMCO's concession would apply to those areas granted to Saudi Arabia under an international agreement. The Department's summary of the conversation with Wood on this latter question reads as follows:

Mr. Wood . . . emphasized that he did not feel that this second
problem was one which involved the Department of State.
Mr. Henderson agreed that this was so as regarded the matter of
interpretation of the Aramco concession, but he added that it
would be a most unfortunate development should a major dispute
arise between American companies interested in the Persian Gulf.
Such a dispute would be embarrassing to the United States in view
of various political factors affecting the entire Middle East.[10]

How competition between two American companies for concessions
in the same region could be 'embarrassing to the United States', the
world's foremost advocate of a competitive economy, Mr Henderson
did not say. In any event the State Department avoided this
'embarrassment' by giving informal backing to ARAMCO.

In a meeting on 16 June 1948, one day before Henderson assured
Superior officials that the 'Department has endeavored to treat all
companies alike', Childs, the US Minister in Saudi Arabia, had a
meeting with Shaikh Yusuf Yassin, Deputy Foreign Minister of Saudi
Arabia, in which they discussed the ARAMCO-Superior dispute.
The Saudi Government had received a favourable offer from Superior
for off-shore concessions. It included payment to the government
of a minimum of 50 per cent of net profits, rising on a sliding scale
to 75 per cent, as well as the appointment of a Saudi general manager,
who would have equal status and salary with the foreign general
manager. As Childs pointed out, the appointment of a Saudi general
manager

has made a very special appeal to the Saudi Arabian Government
in view of the strong feeling that has existed for some time in the
minds of Saudi Arabians against what they regard as discrimination
on the part of Aramco against Saudi Arabians in the provision
of living accommodations and social and economic equality of
treatment.[11]

Upon receiving this offer, the Saudi Government gave ARAMCO
until 30 June 1948 to make a counter-offer. In reply ARAMCO
requested a delay until 31 July. Shaikh Yusuf asked Childs for
advice on how to handle the matter. In a 'secret' despatch to the
State Department, Childs recorded his reply:

I stated to Shaikh Yusuf that I felt this to be a matter between the

Saudi Arabian Government and Aramco with which the United
States was not officially concerned. He pressed me for my personal
view.

    I stated that speaking entirely personally I would answer him on
the basis of a homely analogy. If I had been doing business with a
firm or individual over a period of some years, and that firm or
individual asked me for time in which to consider a problem
arising through the interposition of a third, and hitherto unknown
individual or firm, I would feel that as a matter of courtesy, if
nothing else, the individual or firm with whom I had been doing
business was entitled to obtain the delay desired, particularly if
no prejudice was done thereby to either party.

    Shaikh Yusuf nodded and appeared to accept my reply to
his question as a reasonable one.[12]

Perhaps it is not surprising that Childs, who had previously praised
ARAMCO for its 'very high sense of social responsibility',[13] should,
on his own initiative, violate the proclaimed US policy of maintaining
neutrality between competing US companies. What is surprising
however is that he felt secure in reporting this violation to his
superiors. What is even more surprising is that his superiors took
no action against him.

    Whatever were the real reasons for ARAMCO's request for a
delay is a matter of speculation. Any delay, for whatever reason,
was bound to work to ARAMCO's advantage. Superior did not
have the financial resources of ARAMCO. It could not afford to wait
as long as ARAMCO for a return on its investments. Unless Childs
was exceptionally naïve, he must have known that any delay would
have had a much greater financial impact on Superior than on
ARAMCO. By suggesting that the Saudis accede to ARAMCO's
request, he was in effect helping ARAMCO raise the cost of entry for
Superior.

    His superiors also helped delay a final decision. On 21 July 1948,
three executives of Superior held further discussions with State
Department officials in Washington. The Superior group claimed that
the Saudis were ready to grant Superior the concession provided that
the US Government assured the Saudis that the area came under
Saudi jurisdiction and that the US Government would not interfere
with the granting of a concession to a firm other than ARAMCO.

    The Department refused to grant these assurances on the grounds
that 'it would be unwise and dangerous for any concession to be

granted covering the submerged area of the Persian Gulf until some
orderly division of the submerged area as between the riparian states
had been effected'.[14] When Wood, of Superior, pointed out that the
concession was in an area not subject to conflicting claims from
bordering states and therefore there was no reason why the granting of
the concession should be delayed, Satterthwaite, of the State
Department, insisted otherwise. According to the Department's
memorandum of the meeting, Satterthwaite stated:

> the Department by all means desired that the final orderly division
> of the submerged area be awaited before any concession be granted.
> It was especially important to do this because if individual rulers
> were to start issuing concessions before their title to do so had been
> made clear the way would be left open for the Russians to enter the
> Gulf and start drilling anywhere they chose.[15]

When Allen, of Superior, noted that the Soviets could be excluded
by either a unilateral declaration of sovereignty over a concession
area or an agreement among the states, Satterthwaite replied that an
agreement 'was preferable in that it would provide a more orderly
approach to a peaceful solution of the problem'.[16]

The Department's refusal to accede to Superior's request turned
out to be particularly fruitful for ARAMCO. While negotiations
between the interested states on riparian rights were going on,
ARAMCO sent a delegation to Saudi Arabia with a text of a
proclamation on off-shore mineral rights. This proclamation was
adopted unilaterally by the Saudi Government on 29 May 1949.[17]
Needless to say, ARAMCO retained its rights to the off-shore areas.
ARAMCO's actions furthermore evoked no apparent protest from
the State Department. When ARAMCO executives met with State
Department officials on 17 June 1949 to report on their role in the
Saudi proclamation, no one in the Department seemed at all concerned
with the effects of ARAMCO's behaviour on other states and oil
companies in the region. Whether by accident or design, the State
Department's actions effectively blocked the entry of Superior into
Saudi Arabia. The 'open door' policy apparently did not extend to
those American companies which did not meet with the Department's
approval.

Afghanistan. In addition to blocking entry the State Department
attempted to steer companies to certain specific areas. In January

1948 Charles Rayner, Washington representative of the American
Independent Oil Company, a firm owned by approximately twelve
independent US companies and controlled by Phillips Petroleum and
Signal,[18] visited Loy Henderson at the State Department to discuss
Rayner's proposed trip to the Middle East. Rayner had previously
been the State Department's Petroleum Adviser. Rayner enquired about
the feasibility of going to Iran, which was then the exclusive domain
of Anglo-Iranian. Henderson, according to the Department's
memorandum of the meeting, told him that 'such a trip by an
American oil man did not now seem desirable'.[19] Henderson
presumably feared that such a trip might upset the negotiations with
the British over riparian rights.

   Henderson did however suggest an interesting alternative to
Rayner — Afghanistan. In view of events in that country some thirty
years later, Rayner's reply turned out to be quite revealing. According
to the State Department, Rayner stated:

> that country was too inaccessible to justify exports and that the
> domestic market in Afghanistan was at present very restricted.
> Furthermore the local terrain was difficult and the proximity
> of the Russians was a constant deterrent to American investment
> in Afghanistan.[20]

One can only speculate whether, some thirty years later when the
United States was desperately trying to find new sources of oil to
replace its losses in Iran, it attempted to open up Afghanistan to
American investment through reducing Soviet influence in that
country and that the Soviets reacted to this attempt by intervening
in Afghanistan. Whatever the truth may be, it is certainly worth
noting that the United States has been interested in exploring in
Afghanistan since the end of World War II.

   Henderson's suggestion indicates how the State Department
tried to direct the smaller US companies to specific, and often less
desirable, areas. While this direction may have been perfectly rational
from the point of view of ensuring entry by American companies
into all parts of the world, it had the effect of protecting the
international oil companies from competition from newcomers in the
area in which the internationals had previously gained a dominant
position. The Department's role was that of a cartel 'secretariat',
controlling entry so as not to disturb the equilibrium achieved by the
cartel.

Growth of Foreign Reserves

The State Department however was not able to exercise complete control over entry. Because of specific economic conditions in the US many smaller firms sought to obtain foreign oil reserves on their own. In addition, the tax legislation existed at that time gave them an incentive to invest abroad.

In the years following World War II the domestic industry reaped enormous profits. The rapid expansion of the economy especially benefited the oil industry. This expansion was both motivated by and resulted in a large increase in motor car ownership. Between 1946 and 1953, for instance, automobile registrations rose from 34.4 million to 56.3 million,[21] an increase of 64 per cent. Gasoline consumption consequently soared from 30 US trillion gallons per year to 49 US trillion[22] or at an annual rate of more than 7 per cent.

In response to this increasing demand, petroleum prices climbed much more sharply than those of other commodities. The wholesale price of petroleum and its products advanced by 70 per cent while those of all other commodities rose by only 40 per cent.[23] As a result, the flow of profits to the refining industry, as reported by the Federal Trade Commission and the Securities and Exchange Commision, was higher than in any other manufacturing industry in every year from 1947 to 1953. In these seven years the refining industry amassed $13 billion in after-tax net profits. Its profits accounted for one-sixth of the total profits of all manufacturing industries.[24] The depletion allowance provision of the income tax, which allowed firms producing crude oil to deduct 27½ per cent of the value of their crude oil sales from their taxable income, further enhanced the flow of funds to the industry.

Initially, a sizable portion of these funds was invested in the domestic industry. Proved reserves consequently rose by 38 per cent between 1946 and 1953.[25] Productive capacity, i.e. the capacity to produce oil from existing wells, jumped by approximately 50 per cent.[26] The quantity demanded, as measured by domestic production plus imports plus changes in stocks, however, increased at a much slower rate, by only 33 per cent.[27] This relatively slow rate of growth in demand placed downward pressure on domestic oil prices.

Domestic Production Controls. The authorities in Texas and other oil-producing states countered this downward pressure by restricting production. Since the 1930s the US Government had permitted the oil-producing states to restrict production through the imposition of 'market demand prorationing'. Officially justified as a 'conservation' measure, this type of 'prorationing' turned out to be a price maintenance system. As practised in Texas and most of the other oil states, the system empowered a regulatory authority to place a ceiling on the amount of oil that could be produced each month. This ceiling was determined essentially by the 'nominations' made by refiners to the regulatory agency. In Texas, for instance, the Texas Railroad Commission, that state's regulatory authority, required all refiners intending to purchase oil in the state to notify it about the quantities they wanted to purchase. The Commission then restricted output to these quantities.

These restrictions prevented a 'surplus' of oil from being produced, 'surplus' being defined as the production of more than is demanded at the prevailing price. The prevention of this 'surplus' had the intended effect of either reducing or completely eliminating downward pressure on prices. In order to prevent total production from exceeding refiners' 'nominations', the regulatory agency assigned each well a production quota, called an 'allowable'. Any well-owner who exceeded his 'allowable' faced severe economic and legal sanctions. The prorationing system thus cartelised the US domestic oil industry, with the state regulatory bodies playing the role of a cartel 'secretariat' through their setting of production ceilings and their assignment of 'allowables'.

During the 1950s the state authorities started cutting back 'allowables' significantly. While these cut-backs maintained or even increased crude prices, they began to threaten the profit potential of many wells, since producers could not sell as much oil as before. This reduction in profit potential led many oil companies to look for reserves abroad, where, with the exception of Canada, prorationing did not exist. Thus, instead of investing their profits in the United States, they sent them outside the country in an effort to keep their future profits from falling.

The depletion allowance gave integrated refining companies, i.e. those companies owning both refineries and producing wells, a further incentive to invest abroad. The reductions of 'allowables' meant that many refining companies were not permitted to produce from their own wells as much oil as they required. They were thus forced to buy oil from other producers. When they did this, they both lost the benefit of the depletion allowance, which could be claimed only by sellers of crude oil, and gave this advantage to their competitors. By producing abroad they could take full advantage of this allowance. In addition they could, as has already been pointed out, credit their income tax payments to foreign governments against their US income tax obligations. If they invested in the Western Hemisphere, they also received a 14 percentage point reduction in their income tax rate.[28]

New Entrants Abroad. Given these economic and tax incentives, many firms entered and some, like Standard of Indiana, re-entered the international arena. Fanning estimated that 162 US firms moved overseas for the first time between 1945 and 1958.[29] Many were successful in finding oil in a number of countries. These countries included the Neutral Zone between Kuwait and Saudi Arabia, where concessions were granted to the American Independent Oil Company and the Getty Oil Company.[30] Those two firms entered the Neutral Zone after receiving State Department backing. They were examples of controlled entry.

In other parts of the world, and especially in Latin America, entry was relatively uncontrolled. Smaller American firms found oil in such diverse countries as Venezuela (which in 1956 reversed its 1945 decree prohibiting the granting of new concessions by opening up its territories to the internationals and newcomers alike[31]), Libya, Nigeria, Australia, Indonesia, Peru and Ecuador.[32] By the end of the 1950s these newcomers, who had virtually no foreign holdings at the close of World War II, owned approximately 30 per cent of the overseas reserves of American companies.[33] In Venezuela alone they found almost 3 billion barrels of reserves by 1962 and produced some 400,000 barrels per day during that year.[34] The result of all this activity was the creation of a world-wide 'surplus' of oil, i.e. a 'surplus' in the sense that far more oil could be produced than could be absorbed by the market at the prices then prevailing.

The Surplus of Oil

This surplus had the potential of creating both opportunities and problems for the international oil companies and the US Government. As long as the surplus remained under their control, they could allocate world oil production as they desired. This allocative power was especially important to US policy planners, who, as was mentioned in chapter 5, wanted to supply Europe with Eastern Hemisphere, and particularly with Middle Eastern, oil and to supply the United States with Western Hemisphere oil. Given a surplus in both Hemispheres, such a division of production was possible. US policy attempted to bring about this division.

Surplus as a Means of Control

The international oil companies, regardless of nationality, welcomed the surplus as a means of enhancing their bargaining power *vis-à-vis* the host countries. As long as there was a surplus, the companies could threaten to reduce production in a recalcitrant country by increasing production in those countries whose governments were more 'reasonable'. The surplus was the means through which the international companies were able to skim off a high portion of the economic rents produced by oil.

Iran under Mossadegh was an example of how a surplus can be used against a host country. Blair points out that despite the 'virtually complete loss of Iran's output' during the years 1951-4, production in the OPEC countries rose at a steady rate of 9.55 per cent per year, i.e. at precisely the same rate as production had risen in the entire period from 1950 to 1972.[35] The shortfall in Iranian production was more than offset by substantial increases in production in Kuwait, Saudi Arabia and Venezuela.[36] If this surplus capacity had not existed, the world would have been forced to come to terms with Mossadegh. Probably more than any other single factor, this surplus was responsible for the overthrow of Mossadegh. At that time OPEC's reserves and production were under the complete control of the international oil companies. This control enabled them to win in Iran.

The US Government, for its part, welcomed a surplus in that, aside from purely strategic reasons, it enabled the government to resist demands by the governments of host countries. A 1945 State Department memo, referring to demands by the Saudi Government, warned that it 'is important that no precedent be created which might lead governments of other countries in which American companies control strategic raw materials to expect that they can count upon US Government assistance in meeting their budgetary difficulties'.[37] The greater the surplus the less vulnerable was the United States to these demands.

Problems of a Surplus

If all this surplus were under the control of the US multinational companies, it would have created no problems. There was a coincidence of interests between these companies and the US Government on the use of the surplus. The new entrants however had interests of their own which clashed with those of the internationals and did not fit into the plans of the government for the geographical allocation of production. These newcomers, eager to make a quick return on their newly-found reserves, stepped up production significantly, shipping most of this increased production to the United States. This inflow into the States threatened the structure of the domestic industry.

Growth of US Imports. In the early post-war years the internationals controlled the bulk of crude imports. As Table 6.1 shows, they accounted for three-fourths of such imports in 1948. With US output

Table 6.1: Internationals'[a] Share of US Crude Oil Imports, Selected Years

Year	Total imports (000 barrels per day)	Internationals' imports (000 barrels per day)	Internationals' share (%)
1948[b]	656	444	74
1954[c]	953	433	68
1959[d]	1,119	418	37
Per cent change			
1948-59	+71	−6	

Notes: a. The international companies were: Gulf, Shell, Socony, Standard (New Jersey), Standard of California and Texaco. Shell was not an American company. b. US House of Representatives, Select Committee on Small Business, *Hearings. Effects of Foreign Oil Imports on Domestic Producers*, 82nd Congress, 2nd Session (US Government Printing Office, Washington, 1950), Part 2, p.75. c. US Special (Cabinet) Committee to Investigate Crude Oil Imports, 'Petroleum Imports, Report to the President of the United States', July 29, 1957, Tables II & III. d. US Oil Import Administration, *Monthly Reports*.

in the chief producing states limited by prorationing laws to 'market demand', any increase in imports compelled various state agencies to cut back domestic production in order to maintain the current price. Thus, by directly controlling the flow of imports, the internationals indirectly controlled domestic production. Since they had heavy investments in US production, they had a large financial interest in maintaining the 'proper' balance between imports and domestic production.

This system also enabled the internationals to maintain their position in the domestic refining market. As long as they were able to control the bulk of imports, they, in effect, controlled the supply going to independent refiners. (Independent refiners are those who do not own crude reserves; they are therefore forced to buy crude in the open market.) By increasing imports and thereby forcing a cut-back in domestic production, the internationals were in a position to compel the independent refiners to buy crude from them since the internationals' imports filled the vacuum brought on by the decline in domestic production.

Because of the industry's price structure, most of the profits in oil are made in the production of crude. The internationals have traditionally pursued a policy of keeping the price of crude high relative to that of refined products. The narrow spread between the price of crude and the price of refined products has brought low profit margins to refining. These low margins have discouraged entry by independent companies into refining. In addition to discouraging entry into refining, the relatively high price of crude increased the value of the depletion allowance available to the sellers of crude. The existing independent refiners who were being forced at that time to buy an increasing portion of their crude from the internationals were in reality financing their competitors in refining. This situation was somewhat analogous to the rebates Rockefeller received from the railroads hauling his competitors' products.

After the newcomers, including many independent refiners seeking their own sources of crude, entered the overseas field, the internationals lost control over imports. As can be seen in Table 6.1, the internationals' share of imports fell from 74 per cent in 1948 to 68 per cent in 1954 and to 37 per cent in 1959. Furthermore, all the increases in imports between 1948 and 1959 came from the newcomers, whose total imports rose by 70 per cent. The imports of the internationals actually declined by 6 per cent during those years.

This precipitous decline threatened to undermine the internationals' ability to control domestic production and, with it, the whole price structure which made crude production much more profitable than refining. It also gave independent refiners the opportunity to purchase imported crude from independent producers, thereby ending the system whereby the refiners financed their competitors. In addition, it opened up the possibility that these independent refiners might receive their own, relatively cheap foreign crude, thereby reducing their crude input costs. Since a reduction in crude costs would raise refining profits, the independent refiners would be able to compete more effectively with the internationals. Increased profits, furthermore, would attract new entrants into refining, thus undermining the internationals' dominant position in the field.

The internationals were not the only group disturbed by the flow of imports into the United States. The independent domestic producers, i.e. those companies whose producing properties were entirely located in the States and who owned no refineries, were also adversely affected by them. As the flood of imports rose, the state regulatory agencies drastically cut back their 'allowables'. The Texas Railroad Commission, for instance, reduced the allowables from 366 days in 1948 to 194 days in 1954.[38] This meant that a well which had been allowed to produce at full capacity for every day of the year in 1948 could produce at full capacity for only 194 days in 1954. It had to be shut-in for the remainder of the year. Because of this policy of reducing domestic production to make room for imports, imports, as a percentage of domestic production, rose from 6 per cent in 1948 to 14 per cent in 1958.[39]

This shutting-in of US production conformed to the policy proposed by Bullitt of keeping US oil in the ground as a strategic reserve with current needs being filled by imports. Because of the political difficulties involved in pursuing such a course, this policy was never adopted. The State Department apparently accepted Collado's plan of allocating Middle East oil to Europe, replacing Western Hemisphere oil there, and supplying Western Hemisphere markets from Western Hemisphere sources. Such a policy would ease the drain on Western Hemisphere reserves without, at the same time, crippling the domestic industry economically.

The actual pattern of imports did not however conform to Collado's plan. Middle East oil began to displace Western Hemisphere oil in the United States as well as in Europe. In 1950, for instance, 59 per cent of US imports of crude and unfinished oils came from Venezuela, the chief source of supply in the Western Hemisphere. Only 23 per cent originated in the Middle East. By 1958, Venezuela's share had fallen to 51 per cent while that of the Middle East had climbed to 31 per cent.[40]

The adverse effect of these Middle East imports on the Western Hemisphere oil industry concerned the US Government. It feared that these Middle East incursions would discourage exploration and development within the Western Hemisphere, a region, which, unlike the Middle East, could easily be controlled by the US military. A decline in exploration and development in this area would lead to a fall in the quantity of new reserves found there and might reduce the surplus under US control. The US attitude towards this area was perhaps best exemplified by Dr Arthur Fleming, director of the Office of Defense Mobilization, who stated in 1955 that 'it has always been the policy in Government . . . to consider these countries [Venezuela and Canada] and others in this hemisphere as within the US orbit when dealing with defense questions'.[41]

Import Control Programme. Thus, the government, as well as the internationals, had an interest in controlling the flow of imports; President Eisenhower, accordingly, issued a proclamation in 1959 imposing mandatory controls on oil imports into the United States.[42] This restrictive measure was imposed at the very same time the United States was actively espousing trade 'liberalisation' throughout the world. It was justified on the grounds of 'national security'. Imports were presumably affecting national security through their adverse impact on domestic oil exploration and development.

These import controls had important repercussions both domestically and internationally. Domestically they strengthened the position of the internationals. First of all, they eased the downward pressures on domestic crude prices, thereby reducing the opportunities for independent domestic refiners to expand their output and for newcomers to enter refining. The squeeze on the independent refiners was mitigated somewhat by an oil exchange programme, which allowed these refiners some access to foreign crude. Nevertheless, this access was not as large as it would have been in the absence of controls. Furthermore, if the price of domestic crude had been allowed to fall, all independent domestic refiners, including those who, because of their inland geographical location, could not have profitably imported, would have benefited from this reduction in their input costs.

Second, the controls enabled the internationals to increase their share of imports. After declining precipitously between 1948 and 1959, the first year of mandatory import controls, their share first stabilised and then began to rise until 1967, when the outbreak of the third Arab-Israeli War changed the world oil picture dramatically. As can be seen in Table 6.2, the internationals' share rose from 37 per cent in 1959 to 41 per cent in 1967. The quantities imported by them increased by 36 per cent; those by the other companies, by less than 25 per cent, reversing completely the pre-import-control pattern.

The improvement in the internationals' position was the direct result of the import quota allocation policy adopted by the US Government. Under that policy, licences to import crude were granted only to those companies having refining facilities in the United States. At the onset of the mandatory control programme, the six international companies operating in the United States controlled more than 40 per cent of the nation's refining capacity.[43] They were thus the chief beneficiaries of this policy which limited quota allocations to those firms with refining capacity. This bias towards the internationals was reinforced by allowing companies to receive quotas based on their 'historical' imports. Since the internationals were the only significant 'historical' importers, they received the main benefits from this provision.

Table 6.2: Internationals' Share of US Crude Oil Imports, 1959-1967

Year	Total imports (000 barrels per day)	Internationals' imports (000 barrels per day)	Internationals' share (%)
1959	1,119	418	37
1960	1,151	435	37
1961	1,201	478	40
1962	1,303	532	41
1963	1,325	523	39
1964	1,404	561	40
1965	1,472	581	39
1966	1,485	562	38
1967	1,398	569	41
Per cent change, 1959-67	+25	+36	

Source: US Oil Import Administration, *News Releases.*

Another aspect of the policy of limiting import licences to companies with refining capacity was that it froze out of the US market independent producers who had acquired foreign oil reserves. Since, by definition, independent producers are non-integrated companies, without refining or marketing facilities, they owned no refining capacity in the United States and they were therefore not entitled to a quota allocation. The lack of an allocation made it extremely difficult for them to sell their oil in the US market. As a result, many independents were forced to sell their overseas reserves to firms with US refining facilities, namely the internationals. The main victim of this provision was Superior Oil, the very same company that the US State Department helped freeze out of the Middle East.

Superior, a non-integrated producer, began production in Venezuela in October 1957, on a concession it received a year earlier. It was a newcomer in that country and had little chance to cement firm marketing arrangements with independent refiners, its main customers, before mandatory import controls were imposed in 1959. As a result,

it had to shut-in 28 of its 59 wells. Finally, in 1964, it sold its
Venezuelan properties to Texaco, which had both an import quota
and refineries overseas. Texaco owned a 235,000 barrels per day
refinery in nearby Trinidad.[44] Thus the quota allocation system helped
the internationals to obtain a greater share of overseas reserves.

It also helped them *vis-à-vis* those previously independent refiners
in the States who had acquired foreign oil reserves. Though these
companies did receive a quota, they could not obtain all the oil they
needed from their overseas sources. In addition, they had a difficult
time marketing their overseas production in excess of their quota
allocation. The internationals, on the other hand, had the option of
sending their surplus production to their overseas refining facilities.
The impact of import quotas on them was not nearly as great as on
the independent refiners.

Another provision of the programme, ostensibly designed to aid
the independent refiners, turned out to be of considerable help to
the internationals. This was the requirement that an import quota be
granted to every refining company in the United States, including
those who, because of their geographical location, would never import
oil. In order to realise the value of this quota, they would exchange it,
on a barter basis, with coastal refiners, who were in a position to
utilise imported crude.

These quotas had a value because the restrictions on imports raised
the price of domestic crude above that of foreign crude. If there had
been no restrictions, the price of domestic crude could not have
exceeded the foreign price plus transportation and handling costs. The
mandatory control programme enabled the domestic price to pierce the
ceiling which would have been imposed by unrestricted imports. The
differential between the domestic and foreign price was estimated by
President Kennedy's Petroleum Study Committee to have been $1
per barrel[45] and by Professor M. A. Adelman to have been $1.25[46]
at a time when the domestic price was about $3 per barrel. By trading
its import quota to a coastal refinery, an inland refiner could receive
a portion of its domestic crude at the import price.[47]

These exchanges provided an incentive for the establishment of close
links between certain coastal and inland refiners, which could be
interpreted as anti-competitive behaviour, and for mergers between
refining companies. By buying out an inland refiner, a company with
coastal refineries would also obtain that refiner's import quotas. All
told, the entire system of assigning quotas led to a rash of mergers,
which significantly strengthened the position of the internationals.

Among the mergers which probably were designed to take advantage of the programme's quota allocation system were: Gulf Oil's acquisition of the Wiltshire Oil Co of Los Angeles and most of Cities Service's mid-continent refining and marketing facilities; Standard Oil's (New Jersey) acquisition of Tidewater's refinery in Avon, California; Texaco's acquisition of Paragon Oil and the White Fuel Company; Standard Oil of California's purchase of Standard Oil (Kentucky); and Shell's purchase of El Paso Natural Gas Company's refining and marketing facilities and its acquisition of Buckley Brothers, a small New England oil firm.[48]

The oil import programme, in addition to strengthening the position of the international oil companies, also favoured Western Hemisphere oil over that from the Middle East. As Table 6.3 shows, the Western Hemisphere share of imports remained stable, at 70 per cent, during the period of import controls while the Middle East share fell from 20 per cent in 1958, the last year before the imposition of mandatory controls, to 8 per cent in 1967. Though the Middle East share in that year was affected by the Arab-Israeli war and the subsequent blockage of the Suez Canal, it had been declining steadily since 1958. The war merely accelerated this decline. The remainder of the imports came from such diverse areas as Africa, Indonesia and several Latin American countries. This geographical pattern of imports was achieved through (1) the imposition of less stringent controls on the imports of residual fuel oil and (2) the exemption of Canadian crude imports from all controls.

Residual fuel oil is one of the chief products of a refinery, the others being gasoline, kerosene and the middle distillates. Of all these products, the demand for residual is the most elastic, i.e. the most responsive to small changes in price. The relatively high elasticity of demand for residual is due to the availability of close substitutes, namely natural gas and steam coal. Residual, unlike the other products, is used mainly as an under boiler fuel for heating liquids in thermal electric power plants and in large industrial establishments. Since this heating function can also be performed by natural gas and coal, many utilities and industrial plants have boilers which can switch to these alternative fuels. A slight change in the relative price of these fuels can thus induce a large shift in usage. This price sensitivity has meant that the price of residual fuel oil, on a calorific basis, cannot be higher than the prices of natural gas and steam coal in the relevant markets. Because of US Government regulation of gas prices and a declining demand for coal, the prices of these products were low. Refiners wishing to market

Table 6.3: Geographical Origins of US Imports of Crude Oil and Products, 1958-1967 (per cent of total imports)

Year	Western Hemisphere Venezuela and Dutch West Indies[a]	Canada	Total Western Hemisphere	Middle East	Other[b]
1958	64.7	4.9	69.6	20.3	10.2
1959	65.4	5.3	70.7	19.2	10.1
1960	65.6	6.4	72.0	17.6	10.4
1961	60.9	9.7	70.6	17.8	11.6
1962	60.8	11.7	72.5	14.8	12.7
1963	59.2	11.8	71.0	13.6	15.4
1964	58.2	12.6	70.8	13.4	15.8
1965	57.1	12.4	69.5	13.8	16.7
1966	54.9	15.7	70.6	12.8	16.6
1967	52.7	18.7	71.4	8.3	20.3

Notes: a. Includes products refined at Aruba and Curaçao from Venezuelan crude.
b. Includes Africa, Indonesia and some Latin American countries.
Sources: US Bureau of Mines and US Oil Import Administration

residual had to price within the ceiling imposed by these fuels.

No such restrictions applied to the other refinery products, whose demand elasticity, at least in the short-run, was much lower than that of residual fuel oil. As a consequence, refiners charged a higher price for these products than for residual. The ensuing product price structure gave refiners an incentive to reduce the residual fuel oil yield from a barrel of crude and to increase the yield of the higher valued lighter products, and especially that of gasoline. Technological innovations, like the cracking process, made such a change in yields possible. As a result, refiners reduced their output of residual fuel oil. Along the East Coast of the United States, where the demand for residual fuel oil was strongest, the output of residual fell from 242,000 barrels per day in 1951 to 191,000 barrels in 1958. This decline was more than offset by an increase in residual imports from 326,000 to 489,000 barrels per day during the same period.[49]

The residual came mainly from refineries in Venezuela and in the Dutch West Indies' islands of Aruba and Curaçao. These refineries used the heavy crude oil of Lake Maracaibo as an input. Because this heavy crude yields a high percentage of residual fuel oil, its price is lower than the lighter crudes, which yield a relatively high percentage of gasoline. The oil companies hence found it more profitable to produce their residual from the low-priced Venezuelan heavy crudes than from the higher-priced American and Middle East lighter crudes.

In principle they could have refined the heavy Venezuelan crude in the United States, especially since it is normally cheaper to ship crude long distances than to ship products. Venezuelan policy however forced them to locate refineries in Venezuela. Because of this policy refining capacity in Venezuela expanded from 285,000 barrels per day in 1951 to 820,000 in 1959.[50] Faced with this forced expansion of capacity, the oil companies quite logically decided to use it to convert their heavy crude into residual fuel for sale in the United States while at the same time they reduced the unprofitable residual output of their US refineries. Thus, in the period between 1951 and 1958, the internationals who had refineries both on the East Coast of the United States and in the Venezuelan-Dutch West Indies complex[51] increased, as Table 6.4 indicates, their imports of residual fuel by 50 per cent. The imports of crude oil from the Venezuelan-Dutch West Indies region, however, rose at a faster rate, by 95 per cent.

Table 6.4: Per Cent Changes in US Imports of Crude Oil and Residual Fuel from Venezuelan-Dutch West Indies Region, 1951-1958 and 1958-1967

Years	Crude Oil	Fuel Residual Oil
1951-8	95	50
1958-67	−2	115

Sources: US Bureau of Mines and US Oil Import Administration

After the imposition of mandatory import controls in 1959, the quantity of crude oil imported from Venezuela actually fell. By 1967 it was 2 per cent below the 1958 level. Residual imports from the area, on the other hand, skyrocketed. The less stringent restrictions on imports of residual induced its substitution for crude. This substitution led to some important results.

First, it contributed to the continued dependence of the Atlantic Coast region in general, and of New England in particular, on residual fuel oil. If the restrictions on residual imports had been as severe as those on crude, industries in these regions would have switched to natural gas and coal. Second, it provided Venezuela with a partial offset to the loss in income flowing from import controls. The Venezuelan Government, while recognising this benefit, would have preferred that the United States increase its imports of the higher-valued lighter crudes rather than those of the lower-valued heavy crudes. Third, it benefited the international oil companies with refining facilities in the Venezuelan-Dutch West Indies region by allowing them to recover their investments on the refineries built in the mid-1950s. It also enabled the internationals to increase their control over imports since most newcomers to Venezuela did not have refining facilities in the area. The substitution of product for crude imports thus gave the internationals an important advantage. Fourth, it conformed to the general US policy of directing Western Hemisphere oil to the United States and Middle Eastern oil to Europe. As can be seen in Table 6.3, Venezuela's share of total imports into the United States had fallen between 1958 and 1967. This fall would have been considerably steeper if the import programme had not encouraged the substitution of residual for crude. Finally, it had an adverse impact on the domestic coal industry. Hard hit by the loss of its traditional markets in home-heating and railroads, that industry sought to improve its position through expanding in the fast-growing electricity generation market. The growing imports of residual fuel foreclosed that market along the East Coast, contributing to a drop in coal production and a closing of mines. A similar pattern, with dire consequences for the future, occurred in Europe.

Despite the drop in Venezuela's share, the Western Hemisphere share remained stable because the rise in the Canadian share offset the fall in the Venezuelan. Canada's share rose because the US Government exempted its imports from the control programme. This exemption was granted to Canada under an amendment to the original import proclamation which removed all barriers on imports 'which are transported into the United States by pipeline, rail, or other means of overland transportation from the country where they were produced'.[52]

It is significant that the original proclamation did not provide for such an exemption. It would have seemed logical to grant one in a programme ostensibly based on protecting the natural security. Canadian oil could, after all, provide a safe alternative fuel source in the event of a national emergency. The importation of Canadian oil furthermore conformed to the objective of favouring Western Hemisphere oil over that of the Middle East. Despite the obvious advantages of Canadian oil, the US was reluctant to grant Canada any exemption. The exemption was granted only because the Canadian Government insisted on it and had enough bargaining power to force the US to accede to it.

At that time the importation of Canadian oil did not serve the interests of the internationals. Most of Canada's oil was produced in the Province of Alberta. After the discovery of a major oil field in that province in 1947, Alberta began to expand its oil markets by supplying crude to other parts of Canada, the US West Coast and the US Midwest. This expansion was accelerated during the second Arab-Israeli War of 1956-7. After the end of that war, Alberta's capacity to produce oil greatly exceeded the demand for its oil, which had shrunk because of the restoration of peace in the Middle East and the onset of a recession in the United States.

In order to protect its local producers from this decline in demand, the Alberta Government began to impose prorationing restrictions, similar to those in effect in Texas. The allocation programme was administered in such a way as to force the internationals to bear the brunt of the decline in demand. Under the programme, those international companies wanting to ship more Alberta oil to their refineries were forced to cut back production of their own oil and to buy the required quantities from independent producers. The internationals were naturally reluctant to purchase this relatively high-priced Alberta oil in the market, preferring to use their own. In addition, such purchases would deprive them of the benefits of both the Canadian and American depletion allowances. From their point of view, the importation of Alberta oil was not economically attractive. Those who had previously imported Alberta oil had either drastically reduced or completely eliminated their purchases. At the onset of the mandatory import programme, most Canadian oil went to five small refineries, owned mainly by nonintegrated companies, located in those Midwest states close to the Canadian border.[53] If the internationals had had their way, Canadian imports would not have made any further penetration into the United States.

What made both the international oil companies and the US Government change their minds was the threat by the Canadian Government to build a pipeline from Edmonton, Alberta to Montreal. Montreal, which was Canada's largest refining centre, operated exclusively on imported oil. Four of its five refineries were owned by the international oil companies, who controlled 92 per cent of that city's refining capacity in 1958.[54] The internationals were then supplying the Montreal refineries with their own oil produced mainly in Venezuela.

With the construction of a pipeline from Edmonton to Montreal, the internationals would have been forced to substitute prorationed Alberta crude for their own Venezuelan oil. The internationals resisted such a proposal, which would have significantly reduced the profitability of their Montreal operations. They were backed in

their opposition by the US Government since the proposed pipeline
would have drastically curtailed imports from Venezuela. Because
Canada at that time was Venezuela's second best customer, accounting
for 16 per cent of its export sales,[55] a significant cutback in imports
would have inflicted serious damage on the Venezuelan economy,
especially when that cutback would be added to the effects of the US's
own import restrictions. The United States apparently feared the
political consequences of a major economic downturn in a country
in which US corporations had invested heavily and which was located
in an area which the US had traditionally tried to dominate.

As a result, the US Government offered Canada a compromise. It
agreed to exempt Canadian imports if Canada abandoned the proposed
pipeline. Following this agreement, Canada adopted what was officially
called the NOP (National Oil Policy). Under the NOP the pipeline to
Montreal was not to be built, leaving that area and the entire country
to the east of it totally dependent on imported oil. The rest of Canada,
i.e. the area lying to the west of the Ottawa River all the way to the
Pacific Ocean, was reserved for domestic crude. Though this cartel-like
division of the Canadian market preserved such important cities as
Toronto for domestic producers, it still did not give Alberta a sufficient
market to reduce significantly its shut-in capacity. The NOP excluded
Alberta oil from a region which consumed more than one-third of the
country's oil.

The Canadian Government therefore insisted that the international
oil companies take full advantage of the Canadian exemption by
shipping enough Canadian oil to the United States to compensate
Alberta's producers for the loss of markets in Eastern Canada. The
Royal Commission on Energy (the Borden Commission), established
to study the problem of imports, stated the government's view in the
following words:

Having regard to the international associations of the refiners in
Montreal and in the Maritime provinces and to the large shut-in
capacity of crude oil in Western Canada, in which most of these
refiners have substantial ownership, it is our view that these refiners
should be prepared to strive assiduously to offset their imports of
foreign crude by exports to United States markets. These refiners
should be prepared to work out private commercial arrangements
with their suppliers or affiliated companies or with other companies
which have a large stake in Canadian oil production and could
utilize Canadian crude in United States refineries.[56]

Failure to enter into such arrangements, the Commission warned, would lead to the construction of the pipeline to Montreal and to the licensing of imports into Eastern Canada. The international oil companies apparently took the Commission's threats seriously for, as Table 6.3 shows, they increased their imports of Canadian crude significantly. By 1966 Canada's share of total imports exceeded that of the Middle East.

Forced to increase their purchase of Alberta crude, the internationals turned their attention to changing the province's prorationing system, which, as was mentioned earlier, favoured the small, independent producer. After much debate, in 1965 the province introduced a new allocation system, much more favourable to the internationals.[57] This new system enabled the larger firms to receive a higher share of total allocations. Following this change the operations of the international oil companies in Alberta became more profitable.

The increased penetration of Canadian oil into the US was consistent with the US policy of directing Western Hemisphere oil to US markets. Of all the imported oils, Canadian oil was certainly the most secure. This penetration also was an important step in establishing a 'Continental Energy Policy'. Under such a policy the energy resources of the North American continent were to be pooled. A common market in energy, free of all tariffs, import quotas and other barriers to trade, would be established. Energy would presumably be produced and utilised for the common good of all North Americans.

Needless to say, the chief beneficiary of this type of energy sharing would be the United States, the country less generously endowed with energy resources than either of its neighbours, Mexico and Canada. Not surprisingly, the attempts to adopt a 'Continental Energy Policy' aroused bitter opposition in the latter two countries. In Canada a long and often acrimonious debate has been in progress during the last three decades between the 'Continentalists', who favour energy sharing, and the 'Nationalists', who want to develop the country's energy resources for its own needs. Because of this nationalist opposition, Canada has not yet officially agreed to a continental policy. Nevertheless, it has taken steps which have moved the country closer to such a policy.

The failure to build the pipeline to Montreal was one such step. Not only did it make the country's oil production dependent on United States markets but it also prevented the country from developing a national, integrated oil market. Thus Canada had two separate oil regions: one, the region to the west of the Ottawa Valley, was supplied entirely by domestic crude; the other, the region to the east of the

Ottawa Valley, was supplied entirely by imported crude. The implications of this market division became apparent during the oil embargo of 1973/4 when Eastern Canada suffered oil shortages and Western Canada did not have the transportation facilities to alleviate the shortage in the East. While some oil was shipped from Alberta in tankers from the Canadian Pacific Coast through the Panama Canal to Montreal, it was not nearly enough to meet the East's needs. Such a long, cumbersome journey could never be an effective substitute for a pipeline. Recognising this, the Canadian Government in March 1975 made an agreement with the pipeline company to extend the line to Montreal.[58] This extension, which in 1981 had a capacity of approximately 300,000 barrels per day, supplies Eastern Canada with only one-third of its oil needs. The remainder is supplied by imports.

All told, Canada imports about one-third of its petroleum. Had a pipeline to Montreal been constructed in the 1950s, domestic oil would have provided all, or nearly all, of the nation's oil. This self-sufficiency would not only have protected the country against an oil cut-off but also against the repercussions of higher oil prices. Canada, as a nation with relatively abundant energy supplies, has the potential of insulating itself against the vagaries of the world market. It can, if it so desires, provide its residents with oil at a price below that of the world market.

Canada has, in fact, been pursuing such a low price policy since 1973.[59] The dependence of Eastern Canada on imports has made the implementation of this policy costly. In order to protect the residents of Eastern Canada against the high-priced foreign oil, the federal government has been paying a subsidy of several billion Canadian dollars a year to the international oil companies, who import practically all of Canada's oil. This subsidy enables the companies to buy oil at the higher world price, resell it in the Canadian market at the lower domestic price and still make a profit. As the gap between the domestic and world price of oil widens, the size of the subsidy increases, creating domestic pressures for raising the national price towards the world level.

The debate over oil prices has been a major political issue within Canada. Undoubtedly the fall of the short-lived Progressive Conservative regime of Prime Minister Joe Clark in February 1980 was due in large measure to its attempt to raise prices sharply. Even if Canada were completely insulated from the world market, it would still debate this issue since both the oil companies and the oil-producing province of Alberta have a vested interest in increasing prices. Self sufficiency, however, would have enabled Canada to make a truly independent decision on the level of domestic prices. Presumably that decision would best correspond to its needs and interests. The imports with their concomitant large subsidies make it extremely difficult for Canada to make such an independent decision.

In retrospect, serious doubts can be raised as to whether the adoption of the NOP served Canada's interests. The evidence indicates that the construction of the pipeline to Montreal in the late 1950s would have served her interests better. But there is little doubt that the NOP did serve the interests of both the US Government and the international oil companies. The US Government's interests were served in that Canada did not reduce its imports of Venezuelan oil after the imposition of mandatory import restrictions by the United States in 1959. As Table 6.5 shows, Canada's imports from Venezuela rose by 20 per cent between 1958 and 1965. This increase helped Venezuela cushion the shock of the US mandatory control programme. The NOP thus enabled the US to escape some of the foreign policy consequences of its import restrictions. The cost to Canada was increased dependence on imports at the same time the US was deliberately trying to reduce its own dependence on imports.

Table 6.5: Canadian Imports of Crude Oil from Venezuela and the Middle East, 1958–1965 (000 barrels per day)

Year	Venezuela	Middle East
1958	197	87
1959	204	105
1960	200	138
1961	225	132
1962	236	124
1963	249	143
1964	282	98
1965	236	147
Per cent change, 1958–1965	20	69

Source: Canadian Petroleum Association, *Statistical Year Books*

The international oil companies benefited not only because they could sell their own oil, which was not prorationed, in the Canadian market but also because they could increase their sales of Middle

Eastern oil to Canada. The US control programme restricted their scope for selling this oil in the US. Faced with a mounting surplus of Middle East oil and pressures by the host governments to market that oil, they did not want Canada to impose restrictions similar to those of the United States. The adoption of the NOP enabled them to shift some Middle East oil from the closed American to the open Canadian market. As can be seen in Table 6.5, Canadian imports of Middle East oil grew much more rapidly than those from Venezuela. While this increased dependence on Eastern Hemisphere oil may not have been in conformity with State Department policy, it nevertheless enabled the US-owned international oil companies to dispose of some of their surplus Middle East oil, a surplus that had been exacerbated by the US oil import programme. The US Government presumably preferred this increased Canadian dependence on Middle Eastern oil to the building of the Montreal pipeline and the concomitant adoption of import controls. These increased Canadian imports of Middle East oil were however insufficient to deal with the Middle East surplus. The Canadian market potential, consisting of fewer than 20 million people, was after all small when compared to that of Western Europe. It was only natural therefore that the international companies looked to this area as the main market for their growing surplus of Middle East oil. In this they were actively supported by the US Government.

Notes

1. See for instance Galbraith's theory of 'Countervailing Power' in his *American Capitalism*, Sentry edn (Houghton Mifflin, Boston, 1962), pp.108-34.

2. W. J. Kennedy (ed.), *Secret History of the Oil Companies in the Middle East* (2 vols., Documentary Publications, Salisbury, N.C., 1979), vol. 1, p.111.

3. In his report to the State Department, Childs stated that he 'was not impressed at all by the group from the Superior Oil Company nor was, I may add, the crew of the private plane transporting them. Members of the crew, who have some familiarity with this area, described the party to a member of the Legation staff as a group of joy-riding promoters travelling on a junket at the expense of the company.' See ibid., p.111.

4. Ibid., pp.110-11.

5. Ibid., pp.107-8.

6. Ibid., p.162.

7. Ibid., p.96.

8. Ibid., p.199.

9. A. France, *Le Lys Rouge* (1894), as quoted in J. Bartlett, *Familiar Quotations*, 14th edn (Little, Brown, Boston, 1968), p.802.

10. Kennedy, *Secret History of the Oil Companies in the Middle East*, vol. 1, p.161.

11. Ibid., p.204.

12. Ibid.
13. Ibid., p.155.
14. Ibid., p.217.
15. Ibid., p.219.
16. Ibid., p.220.
17. Ibid., vol. 2, p.382.
18. C. Issawi and M. Yeganeh, *The Economics of Middle Eastern Oil* (Praeger, New York, 1962), pp.178-9.
19. Kennedy, *Secret History of the Oil Companies in the Middle East*, vol. 1, p.139.
20. Ibid., p.138.
21. *Petroleum Facts and Figures: Centennial Edition* (American Petroleum Institute, New York, 1959), p.240.
22. Ibid., pp.246-7.
23. Ibid., p.385.
24. US Department of Commerce, *Business Statistics: 1963 Edition* (US Government Printing Office, Washington, 1963), p.101.
25. *Petroleum Facts and Figures: Centennial Edition*, p.22.
26. *Petroleum Imports, A Report to the Secretary of the Interior* (National Petroleum Council, Washington, 1955), p.51.
27. *Petroleum Facts and Figures: Centennial Edition*, p.213.
28. US Senate, Subcommittee on Multinational Corporations of the Committee on Foreign Relations, Hearings, *Multinational Corporations and United States Foreign Policy*, 93rd Congress, 2nd Session (US Government Printing Office, Washington, 1974), part 4, p.117.
29. L. Fanning, *The Shift of World Petroleum Power Away from the United States* (Gulf Oil Company, Pittsburgh, 1958), p.33.
30. Issawi and Yeganeh, *The Economics of Middle East Oil*, pp.178-9.
31. 'A Review of Venezuelan Oil Activities during 1956', *Venezuela Up-to-Date*, vol. 7 (July-August 1957), p.14.
32. N. Jacoby, *Multinational Oil, A Study in Industrial Dynamics* (Macmillan, New York, 1974), pp.129, 134-5, 138.
33. A. Gols, 'United States Foreign Oil Investments', unpublished PhD thesis, University of Oregon, 1961, p.21.
34. Jacoby, *Multinational Oil.*, p.138.
35. J. Blair, *The Control of Oil* (Pantheon, New York, 1976), p.101.
36. Ibid., p.100.
37. Kennedy, *Secret History of the Oil Companies in the Middle East*, pp.105-6.
38. The Railroad Commission of Texas, *Annual Reports of the Oil and Gas Division.*
39. E. Shaffer, *The Oil Import Program of the United States: An Evaluation* (Praeger, New York, 1968), p.19.
40. Ibid., p.128.
41. *The Oil and Gas Journal*, 7 November 1955, p.71.
42. 'Adjusting Imports of Petroleum and Petroleum Products into the United States', US Presidential Proclamation 3279, March 10, 1959 (24 F.R. 1781).
43. US House of Representatives, Select Committee on Small Business, *Hearings, Small Business Problems Created by Petroleum Imports* (US Government Printing Office, Washington, 1962), 87th Congress, 1st Session, pp.975-91.
44. Shaffer, *The Oil Import Program of the United States*, pp.152-3.
45. US Petroleum Study Committee, 'A Report to the President', 4 Sept. 1962, p.2.
46. M.A. Adelman, 'Efficiency of Resource Use in Crude Petroleum', *The Southern Economic Journal*, vol. 31 (1964), p.103.

47. For a detailed explanation of how the exchange system operated, see R. Manes, 'Import Quotas, Prices and Profits in the Oil Industry', *The Southern Economic Journal*, vol. 30 (1963), pp.13-24.

48. Shaffer, *The Oil Import Program of the United States*, pp.180-2, 205, 208.

49. Ibid., p.89.

50. *Petroleum Facts and Figures: Centennial Edition*, pp.438-9.

51. Four of the internationals — Standard Oil (New Jersey), Gulf, Standard Oil of California and Texaco — controlled approximately 45 per cent of refining capacity on the East Coast and 65 per cent in the Venezuelan-Dutch East Indies area. See Shaffer, *The Oil Import Program of the United States*, p.91.

52. US President, 'Modifying Proclamation No. 3279 of March 10, 1959, Adjusting Imports of Petroleum and Petroleum Products', Proclamation 3290, 20 April 1959 (24 F.R. 3527).

53. Shaffer, *The Oil Import Program of the United States*, pp.111-12.

54. *Petroleum Facts and Figures: 1963 Edition* (American Petroleum Institute, New York, 1963), p.95.

55. *Actividad Petrolera en 1962: Separata del Informe Economico del Banco Central de Venezuela Correspondiente al Ejercicio Annual 1962, Capitulo de Hidrocarbons* (Banco Central de Venezuela, Caracas, n.d.), pp.9-29.

56. Canada, Royal Commission on Energy, *Second Report to His Excellency the Governor General in Council*, July 1959, p.139.

57. S. Thompson, 'Prorationing of Oil in Alberta and Some Economic Implications', unpublished MA thesis, University of Alberta, 1968, p.73.

58. *An Energy Strategy for Canada: Policies for Self-Reliance* (Minister of Energy, Mines and Resources, Ottawa, 1976), p.154.

59. Ibid., p.152.

7 THE CONTROL OF WORLD OIL: THE TRANSFORMATION OF EUROPE

As has already been noted, the US oil companies had been actively seeking markets in Europe since the end of World War II. The US Government, through the Marshall Plan and other aid programmes, supported them fully in their efforts to penetrate the Continent. This support was the key to US post-war oil policy.

Though in line with the general policy of encouraging US business expansion abroad and with the specific policy of finding markets for US oil companies to enable them to keep their Middle East concessions, this support had a more basic aim — the permanent control of Europe. In the immediate post war years, there was a fear in Washington that several European countries, mainly France and Italy, might go communist. Moreover, in such countries as Britain and West Germany, communist-led unions were dominant in the coal fields.[1] Coal, which was then Europe's principal energy source, might thus come under communist control.

Middle East oil, on the other hand, seemed at that time to be firmly under US control. By inducing Europe to switch from coal to oil, the US could keep Europe within its orbit. This strategy was openly bandied about about by US policy-makers in the early post-war years. In his diary, the then US Secretary of Defense, James Forrestal, recorded a conversation that he had with Senator Owen Brewster, chairman of the Senate Commerce Committee, on this subject:

> Brewster said that he had had a long talk with John D. Rockefeller, Jr . . . Europe in the next ten years may shift from a coal to an oil economy, and therefore whoever sits on the valve of Middle East oil may control the destiny of Europe.[2]

The control of Europe was fundamental for US strategy. Though initially motivated by the desire to keep Europe non-communist, it later became necessary to ensure that Europe remained under US hegemony. What the US feared was that an economically independent Europe would pose a serious competitive threat to US enterprises. It was not in the US interests to allow Europe to become strong enough to challenge the US dominance of the world economy. By tying

Europe to US-owned Middle East oil, the US could keep Europe in check. Thus, in the early post-war years, both the US Government and the US oil companies started a campaign to force Europe to become dependent on oil.

Europe's Dependence on Coal

Their task was not an easy one. As can be seen in Table 7.1, solid fuels[3] furnished 85 per cent of Western Europe's fossil fuel[4] requirements in 1950. Oil furnished the remainder. Furthermore,

Table 7.1: Fossil Fuels Consumed in OECD Europe, Selected Post World War II Years (per cent)[a]

Year	Type of fossil fuel		
	Solid[b]	Liquid[c]	Gas[d]
1950	85	15	
1955	77	22	1
1960	64	34	2
1965	47	51	2
1970	30	63	7
1973	23	66	11
1974	24	63	13
1978	22	63	15

Notes: a. The fossil fuels are the hydrocarbons, i.e. coal, oil, natural gas and other fuels created by the photosynthesis process. They are, in fact, stored solar energy. Excluded from this table are such non-fossil fuel energy sources as hydroelectric power, atomic and tidal energy. These latter energy sources constitute only a small fraction of OECD Europe's supply. In 1980 the European members of OECD (Organisation for Economic Corporation and Development) were: Austria, Belgium, Denmark, Finland, France, Federal Republic of Germany, Greece, Iceland, Ireland, Italy, Luxembourg, the Netherlands, Norway, Portugal, Spain, Sweden, Turkey and the United Kingdom. Fuel for overseas marine bunkers is included in the consumption totals. b. The solid fuels are: coal, coke, lignite, peat, wood and the like. c. The liquid fuels are: crude oil, petroleum products and natural gas liquids. d. The gas fuels are: natural gas and town gas.
Sources: 1950, 1955: Organisation for Economic Cooperation and Development, *Energy Policy: Problems and Objectives* (OECD, Paris, 1966), p.150; 1960, 1965, 1970, 1973: *OECD Energy Balances of OECD Countries, 1960/74* (OECD, Paris, 1976), pp.41-6, 51 and 54; 1974, 1978: IEA (International Energy Agency), *Energy Balances of OECD Countries, 1974/1978* (IEA, Paris, 1980), pp.7, 11.

since automobile ownership in Europe was relatively low, the transportation sector did not provide the main market for oil as it did in the United States. In 1960, as Table 7.2 shows, the transportation sector accounted for only 30 per cent of oil consumption in Europe.

Table 7.2: Oil Use by Sector, OECD Europe and United States, 1960 (per cent)

Sector	OECD Europe	United States
Transportation	30	49
(Road)	(23)	(41)
Industry	24	8
(Iron and steel)	(4)	
Other[a]	27	28
Electricity generation	6	3
Marine bunkers	5	4
Manufacturing of town gas	1	
Non-Energy[b]	7	8

Notes: a. Mainly home heating but also includes heating of office buildings, street lighting and agriculture. b. Production of bitumen, lubricants and naphtha. Naphtha is used extensively as a petrochemical feed stock.
Source: OECD, *Energy Balances of OECD Countries, 1960, 1974* (OECD, Paris, 1976), pp.41, 101.

By comparison, it accounted for approximately 50 per cent in the United States.

To rely on the transportation sector to increase Europe's dependency on oil promised to be a long-run process. Aside from an increase through the conversion of the railroads from coal to diesel locomotives (or, alternatively to electric trains, with the electricity being generated by oil-fuelled thermal power stations), an increase in oil demand by the transport sector depended mainly on the automobilisation of Europe, which implied not merely a very significant expansion of private automobile ownership but also the building of ancillary facilities, like highways, service stations and garages. Such a development could not occur overnight.

A much more immediate prospect was the conversion of industry, electricity generation and home heating from coal to oil. Given the proper price incentives, industry could be expected to change its equipment in a relatively short time. That there was considerable scope for such a transformation can be gleaned from a comparison of coal consumption in these sectors in Western Europe and in the United States in 1960. As Table 7.3 shows, coal furnished a far larger portion of the energy requirements of these sectors in Western Europe. For instance, it supplied two-thirds of industry's requirements in Europe as opposed to less than one-third in the US. In electricity it

Table 7.3: Coal's Share of Sectoral Fossil Fuel Energy Requirements, OECD Europe and United States, 1960 (per cent)

Sector	OECD Europe	United States
Industry	64	29
Electricity generation	86	66
Other[a]	67	10

Note: a. Mainly home heating but also includes heating of office buildings, street lighting and agriculture.
Source: OECD, *Energy Balances of OECD Countries, 1960—1974* (OECD, Paris, 1976), pp.41, 101.

accounted for more than 85 per cent in Europe in contrast to 66 per cent in the States. In home heating its portion in Europe was two-thirds while in America it was one-tenth. Because residual fuel oil is competitive with coal, it can make inroads into coal's markets by selling at a price lower than that of coal. By charging a relatively low price for residual, the oil companies were able to drive coal from its industrial and electricity generating markets. As can be seen in Table 7.4, they adopted this pricing pattern in Europe, a pattern similar to that in the United States.

Table 7.4: Petroleum Product Pricing Patterns, North-Western Europe (1951) and United States Gulf (1950) ($ US/barrel)

	Europe	United States
Gasoline	$ 5.19	$ 4.20
Residual fuel oil	2.74	1.65
Difference	2.45	2.55
Refinery acquisition cost of crude oil	3.22	2.72
Oil equivalent price of US coal	5.50[a]	1.91[b]
Refinery margin	0.52	0.66

Notes: a. C.i.f., Europe. b. F.o.b., United States.
Sources: Oil prices and margins, UNECE, *The Price of Oil in Western Europe* (Geneva, 1955), Annex, p.4; coal prices, OEEC (Organisation for European Economic Cooperation), *Europe's Growing Needs of Energy — How Can They Be Met?* (Paris, 1955), p.45.

European Pricing Patterns

The data in this table, taken from estimates made by the UNECE (United Nations Economic Commission for Europe), show that in the early 1950s the price of gasoline both in Europe and in the United States was approximately $2.50 per barrel higher than the price of residual fuel oil. In both places the price of residual was below that of crude. In other words the pricing policies of the oil companies led to one finished product, residual fuel oil, being sold at a lower price than that of the raw material. In both places the price of residual was also lower than the price of American coal, the lowest priced coal on the market.

At first glance, this European price pattern is one that would be expected in a competitive economy. Since Europe had traditionally imported most of its products from the Western Hemisphere, it would seem only natural that the landed prices of these imports would set a ceiling on the prices European refineries could charge. These similar pricing patterns would thus reflect the integration of the world product market.

Upon closer examination, however, this particular explanation does not hold water. First, by the early 1950s, Europe was already becoming a major refining centre. Product imports, and especially those from the Western Hemisphere, were falling dramatically, thereby loosening the links to the US product market. Second, and even more important, was the discrepancy between refinery margins, i.e. the difference between the revenue a refinery receives from selling a barrel of crude and the cost of acquiring that barrel. According to the calculations of the UNECE, these margins were higher in the United States. Competition presumably would set in forces to equalise these margins. But the price structure prevented this equalisation from occurring. This structure did not conform to the differences in product yields between refineries in Europe and in the United States. The refineries in Europe yielded a much higher percentage of residual fuel oil than those in Europe. The reverse pattern held true for gasoline yields. These patterns are shown in Table 7.5.

More than half of the output of European refineries consisted of residual fuel oil, the lowest priced product. In the United States, on the other hand, almost half of the output consisted of gasoline, the highest priced product. Thus the difference in refinery margins.

Under competition, rational behaviour would have dictated that refineries in Europe follow the example of those in the United States by increasing their gasoline yields through the installation of cracking facilities. By using the cracking processes, they could have converted a portion of the residual into gasoline. Residual could have thus served as a substitute for the higher-priced crude in the production

Table 7.5: Petroleum Yield Patterns, North-Western Europe (1951) and United States (1950) (per cent)

	Europe	United States
Gasoline	26	49
Kerosene	2	10
Distillate fuel oil	20	24
Residual fuel oil	51	17

Sources: UNECE, *The Price of Oil in Western Europe* (Geneva, 1955), Annex, p.4.

of gasoline. This use of residual would have significantly raised European refinery margins.

Despite the incentives presented by the price structure, the European refineries, which were mainly owned by the internationals, did not significantly reduce their residual yields. Their production policies were dictated by the consumption patterns in Europe, which differed markedly from those in the United States. These differences are illustrated in Table 7.6.

Table 7.6: Petroleum Production Consumption Patterns, Western Europe and United States, 1953 (per cent)

	Europe	United States
Gasoline	31	51
Kerosene	6	5
Distillate fuel oil	25	21
Residual fuel oil	39	23

Sources: UNECE, *The Price of Oil in Western Europe* (Geneva, 1955), Annex, p.3.

Granted that the refineries followed the European consumption pattern, the question arises, why didn't the European price structure also follow this pattern? In its report, the UNECE noted:

> since the demand for fuel oil relative to that of motor spirit is much higher in western Europe than in the United States, the difference between the prices of these products would be much smaller if prices in Western Europe were determined in the light of the conditions of demand and production ruling there.[5]

Presumably the refineries could have increased their margins by raising the price of residual, the product for which there was the greatest

demand in Europe. But this short-term gain would have endangered
the long-term goal of penetrating the coal market.

The experience of the pre-war years testified to the importance
of this link between fuel oil prices and market penetration. In those
years the price of residual was higher than that of European coal.
This relatively high price limited the market for residual. As a result,
the few refineries that did exist on the Continent were forced to
export their fuel oil.[6] Coal's dominant position remained unshaken.

European Refinery Expansion

The post-war pricing pattern reversed this situation. Within a
short time Europe became a net importer of fuel oil. The expanding
European market served as an inducement to enlarge refinery capacity
and distribution facilities in the region. As a result, the geographical
distribution of US investment in the petroleum industry shifted
decisively towards Europe. As Table 7.7 shows, US oil companies
had placed slightly over one-tenth of their foreign investments in
Europe in 1950. By 1960, this portion rose to approximately
one-sixth. Ten years later, in 1970, it was more than one-fourth.
By 1974, it was almost one-half. By this time Europe had become
the principal area of US oil investments.

The Assault on Coal

Given the relatively low margins received by European refineries,
this heavy investment in refineries appears irrational. But, in reality,
it was a very rational move. First of all, it enabled the oil companies
to increase their sales of Middle East crude. These increased sales
satisfied both the demands of the Middle East governments for
more revenues and the desire of the oil companies for higher profits.
Because of the industry's price structure, profits were made in the
production – the upstream phase – rather than in refining and
distribution – the downstream phases.

It must be remembered that only the independent refineries, i.e.
those who did not own any crude producing properties, paid the full
price for imported crude. This price was based on the posted price in
the Middle East plus transportation and other costs and equalled the
refinery acquisition costs shown in Table 7.4. The actual acquisition

costs of the integrated refiners, i.e. those owned by the international
oil companies, were equal to the tax-paid costs, consisting mainly of
production expenses, tax and royalty payments plus transport and
other charges. Depending on the firms' exact tax and royalty
arrangements and their ability to make use of the US tax credit, the real
acquisition cost of crude was probably no more than half the nominal
cost. Even though the price of residual fuel oil was less than the
reported refinery acquisition cost of crude, the residual price exceeded
the integrated refiners' actual acquisition costs. Thus, from the point
of view of a fully-integrated company, the increased sales of residual
fuel oil turned out to be a profitable operation.

**Table 7.7: Geographical Distribution of US Private Direct Foreign
Investment in the Petroleum Industry, Selected Post World War II
Years (per cent)**

Year	Europe	Latin America	Canada	Other
1950	12	35	12	41
1960	16	27	25	33
1970	28	14	22	37
1974	46	17	27	11
1975	43	13	24	20
1976	44	10	25	21
1977	44	11	25	20
1978	44	11	25	20

Calculated from: *Survey of Current Business*, various issues. Based on year-end
book values. Because of rounding, the sum of some rows may not equal 100.
Note: The shift to Europe was not entirely due to an increase in investment in
refining and marketing facilities. It was also due to investment in North Sea oil.
Furthermore, some of the decline in the Latin American and Other shares
resulted from participation agreements and nationalisations, particularly in
Venezuela and in the Middle East. Nevertheless, the petroleum industry invested
far more in refining and marketing in Europe than in any region outside of
the United States. For the years 1963-76 the industry invested approximately
equal amounts in refining and marketing facilities in the United States and
Europe, about $29 billion in each area. See Chase-Manhattan Bank, *Capital
Investments of the World Petroleum Industry*, various issues.

Second, the oil companies probably viewed their theoretical losses
from reduced refining margins as investments designed to generate
greater future income. To the extent that they succeeded in closing
down coal mines, they would be able to eliminate the restraints on
residual fuel oil prices imposed by coal. Knowing that once coal mines
are closed down they are very difficult and costly to re-open, the oil

companies presumably believed that they could then raise residual prices without fear of competition from coal. In more technical terms, the abandonment of coal mines would lower the price elasticity of demand for fuel oil, thereby significantly increasing the opportunities for raising its price. The companies' fuel oil pricing policy might therefore be viewed as a form of 'predatory-pricing', a policy often practised by Rockefeller by which he sustained short-term losses in order to drive out his competitors.

Third, to the extent that the companies could eliminate coal as a competitor, they would make Europe dependent on their oil, thus giving them a high degree of control over the European economy. This objective coincided with the US foreign policy aim of keeping Europe under US control.

Armand Report

Many Europeans were aware of the implications of the shift to oil. Responding to their concerns, several governments and international agencies conducted studies of and issued reports on energy policy. One of the first of these studies was the Armand Report of 1955. Named after its author, Louis Armand, Chairman of the Board of the SNCF (*Société Nationale des Chemins de Fer Français*), the French national railroad. This report was commissioned by the Council of Ministers of the OEEC (Organisation for European Economic Cooperation), an agency which was an offspring of the Marshall Plan. In his report Armand expressed concern about the inroads foreign oil was making in Europe. 'The most outstanding feature of Europe's situation with regard to oil,' he wrote, 'is undoubtedly the fact that most supplies have to be imported.'[7] He complained that these imports, accounting for 97 per cent of total oil consumption, constituted 'a heavy drain on foreign exchange, a substantial proportion being in hard currency (dollars)'.[8] He advocated instead the development of indigenous energy resources which, he believed, would also reduce energy costs. 'Instead of "more and more power",' he confidently proclaimed, 'the slogan will now have to be "cheap power above all".'[9]

Atomic energy was to be the generator of this cheap power. He gave vent to his unbounded enthusiasm for this new energy source in the following words:

> The first consequence of the dawn of the atomic era must be to abolish the fear of any power shortage in the future; it will be therefore necessary to abandon the idea that power costs in

Europe are linked irretrievably with the price of coal and are therefore bound to rise, as the future costs of atomic energy will certainly be lower than the average prices charged in Europe at present.[10]

He nevertheless recognised that 'atomic energy cannot alone cover the anticipated increase in demand during the next ten years or so'.[11] He therefore conceded that 'a certain proportion of the increased demand for power will have to be met from traditional resources in the next 10 or 20 years'.[12] Coal was the chief 'traditional resource'.

His enthusiasm for coal was surpassed only by his optimism about atomic energy. Though admitting that coal prices will rise in the long-run, he was sanguine about the possibilities of reducing coal costs in the short term, i.e. in the interim before its displacement by cheaper atomic power.

> There can be no immediate prospect of a reduction in coal requirements which are, on the contrary, likely to increase ... Nor should it be forgotten that new equipment will have an important bearing on prices. In these circumstances, the modernisation plans for certain mines will remain just as necessary for many years to come ... No mention has been made of technical developments which ... might lead to substantial progress or even far-reaching changes (e.g. underground gasification ...). In this connection there is considerable unused potential in the European coal and gas industries which, based on lower prices, should be put to minimum use during the next few years until such time as atomic power is in a position to play a more important part.[13]

He also placed high hopes on developing Europe's own oil and natural gas resources, based on natural gas discoveries in Italy and in Southern France.[14] In addition, he regarded the discoveries in Europe's African colonies, and especially those in the French Sahara, as part of Europe's reserves,[15] an attitude not too dissimilar from that of the United States Government to the reserves owned by US companies. Finally, he advocated locating industrial plants near hydro-electric sites in such places as Norway, Austria, Yugoslavia and Africa.[16] Presumably he expected the latter area to remain permanently under European domination.

What is significant about Armand's report was not his specific proposals but his advocacy of European control over energy sources. Most important, he did not look to oil as the means of supplying the Continent with cheap energy. He emphasised instead oil's drain on Europe's limited hard currency reserves.

After examining his report, the OEEC Council decided in 1955 to establish a Commission on Energy and set up a working party to explore the ways and means of cooperating in the development of nuclear energy.[17]

Hartley Report

Shortly after its founding, the Commission on Energy asked Sir Harold Hartley of the United Kingdom to prepare a report on Europe's energy needs. Like Armand, Hartley expressed concern about Europe's growing energy imports. But his worries, unlike Armand's, extended beyond the mere question of the balance of payments. He was also worried about security of supply. Writing in early 1956, before the Arab-Israeli War of that year and the advent of Europe's first oil crisis, he commented that 'there are inevitable risks in the increasing dependence of Western Europe on outside supplies, particularly when most of them must come from one small area of the world'.[18] Though conceding that the 'incentive to substitute oil for coal has been increased by the rise in coal prices in comparison to those of oil',[19] he nevertheless argued that 'coal must remain the mainstay of the Western European energy economy'.[20]

Far less optimistic than Armand about the potential of nuclear energy, he deplored the adverse effects that 'exaggerated statements about nuclear energy'[21] have had on investments in the coal industry. At the same time he conceded that coal could not satisfy all of Europe's energy requirements. He forecast that by 1975 domestic coal production would satisfy slightly less than one-half of Europe's energy needs. The remainder would be supplied by hydro-power, indigenous crude oil and natural gas, and the importation of nuclear fuels, coal and crude oil. He expected that imported crude oil alone would meet between one-fifth and one-third of total energy consumption.[22] (In actuality, crude oil imports in 1975 accounted for over one-half of total energy requirements.)[23] Apparently he regarded this level of crude imports as acceptable from the point of view of security of supply. This acceptance was based on conversations he had with officials of the oil companies. He reported:

The oil companies with whom we had discussions assured us that in view of their present knowledge of oil reserves ample crude would be available to meet the maximum demands of our forecasts (i.e., one-third of energy requirements in 1975). They pointed out that oil is an international industry and that Europe's consumption at present represents 15 per cent of the whole, though this might rise to 20 per cent by 1975.[24]

Despite these optimistic assurances, Hartley did not want oil imports to grow because of anticipated balance-of-payments difficulties. He forecast that the costs of imports would reach $5 billion by 1975, assuming no increase in 1956 price levels, but he added that the prices of these imports 'are certain to rise'.[25] The only way to prevent future serious balance-of-payments difficulties, he argued, was to expand coal production through creating conditions favourable for long-term investments in the industry. These conditions did not exist in 1956. But he was confident 'that given favourable conditions for investment and development on a long-term basis, the coal industry could make a substantial addition to its output'.[26] 'In our opinion', he added, 'Western Europe will never regret its investments in the coal industry.'[27]

To create favourable conditions for investment in coal, Hartley urged that each country in the OEEC adopt an energy policy. While paying lip-service to competition, he clearly implied that policy measures must be substituted for marketplace decisions if Europe were to solve its energy problems.

We think that competition must be allowed to play its full role in giving the consumer maximum choice between energy sources. However, in order to deal effectively with the urgent problems involved in the supply and demand of energy, each Member country will require an energy policy suited to its own circumstances and its needs and resources. This policy should include some measure of coordination between the different forms of energy.[28]

In this he was echoing the view of Armand, who urged the OEEC to avoid 'a situation in which competition between the various forms of energy acts to the detriment of the community as a whole'.[29] It is interesting to note that both Armand and Hartley came to this conclusion despite the oil companies' assurances that Europe had nothing to fear from an increased dependency on imported oil.

Robinson Report

Following the acceptance of the Hartley Report, the OEEC in 1956 established an Energy Advisory Commission, a 'permanent group of high-level experts' under the chairmanship of Professor Austin Robinson of Cambridge University.[30] The nine members of this Commission were to 'act as independent experts to the Organisation and not as official representatives of their countries of origin'.[31] Shortly after its inception, this Commission began its own study of European energy problems.

Completed in 1959, this study proposed solutions radically different from those of Armand and Hartley. Robinson's arguments were based on a presumed 'change from a protracted period of shortage of energy in total to a state of overall surplus'.[32] If there were a danger of a shortage of any fuel, the report continued, 'we would expect coal supplies to be near the upper limit'.[33] Because of this new situation, the Robinson Commission was not 'unduly alarmed at the existence of a "gap" in European energy supplies'.[34] The increasing reliance on imports should pose no special problems because 'there are reasonable prospects that the imports can be financed'.[35]

Dismissing the balance-of-payments concerns of the earlier reports, Robinson looked to the Sahara as a major source of oil imports. Assuming, like Armand, that the Sahara would always remain French, he maintained that these imports would be paid for in French currency and 'corresponding exports may be more than normally easy'.[36] To the extent that the imports would come from other areas, the foreign exchange drain, he claimed, would be more apparent than real. Without giving any quantitative estimates, he concluded that 'the net overseas payment, after allowing for interest and dividends on European holdings in the oil companies concerned, is considerably less than the import costs that appear in the trade returns.'[37] Ignoring the fact that the bulk of the interest and dividends would flow to American, rather than to European, interests, he made an interesting, though probably unwitting, admission, namely that the ability of Europe to pay for its oil imports hinged on the economic and political control of the oil producing areas. In other words, only through a policy of oil imperialism would Europe be able to finance these imports.

Robinson furthermore felt that manufactured exports could also alleviate the balance-of-payments problem. He thus wrote:

if Western Europe can maintain its share of world markets for
manufactures, the import of the increased proportions of the
total supplies of energy that have emerged from our study may
reasonably be expected to be within the probable limits of its
capacity.[38]

The ability to compete in world markets, whether to hold one's share
or to increase it, depends on price effectiveness, which in turn depends
on low input costs. The Robinson Commission therefore emphasised
the cheapness, rather than the security, of energy supplies.

When formulating a long-term energy policy, the paramount
consideration should, in our view, be a plentiful supply of
low-cost energy with a freedom of choice to the consumer . . .
The general interest is best served by placing the least possible
obstacle in the way of economic development of the newer
and cheaper sources of energy.[39]

This emphasis on 'low-cost energy' meant a rejection of Hartley's
suggestion that long-term investment in coal be encouraged. Assuming
that any expansion of coal production would increase costs,
Robinson favoured reducing coal output and concentrating it in those
pits 'where the prospects of increasing productivity are greatest'.[40]
The coal actually marketed should sell 'at a price that will make its
use fully competitive with that of alternative forms of energy'.[41]
Given the residual fuel-oil pricing policies of the oil companies, most
European coal could not be profitably sold at a price competitive with
that of oil. Hence the adoption of a 'low-cost energy' policy spelled
the end of an indigenous one based on coal.

The Robinson Commission did not merely reject coal. It also
down-played nuclear energy. Claiming that 'an active policy of
developing nuclear energy secures a relatively small immediate saving
of foreign exchange at the cost of a very large investment of indigenous
European resources', it suggested that in view of the energy surplus the
'Member countries review their plans for the development of nuclear
energy'.[42] In summary, the Robinson Commission advocated the
abandonment of indigenous energy sources in favour of reliance on
imported oil.

Interestingly, the Commission's report was written after the oil
crisis caused by the Arab-Israeli War of 1956-57. Nevertheless the
Commission was not worried about security of supply. In fact the

Commission stated quite categorically that 'it does not seem likely that shortages of oil supplies will make themselves felt in any acute form by 1975'.[43] These assurances of future oil supplies were based on information supplied by the international oil companies. The Commission consulted these companies in preparing its report.[44]

Because of their implications for the domestic coal industries, Robinson's recommendations received a mixed reception. In his foreword, the Secretary-General of the OEEC was careful to note that 'the views expressed and the conclusions reached in this report represent the opinion of this group of independent experts and thus do not necessarily reflect the views of the Organisation as a whole'.[45] Rather than formally endorse it, the OEEC Council merely voted to publish it.[46]

Nevertheless the governments of most OEEC countries followed the recommendations in varying degrees. So did the ECSC (European Coal and Steel Community). Under rationalisation schemes, coal mines were shut down throughout Europe. Belgium and Holland stopped coal production completely. In Germany and France it was reduced drastically.[47] European coal production peaked in 1957 and has declined steadily since that year.[48] As can be seen in Table 7.8, it had fallen by more than 40 per cent by 1973 while total European energy requirements had more than doubled. Coal production in the relatively prosperous year of 1973 was lower than in the Depression year of 1937 and the early post-war year of 1950.

Coal was displaced by oil, and to a lesser extent by natural gas, in thermal power plants, industrial enterprises, office buildings and households. In all industrial establishments coal consumption fell from 116 mtoe (millions of metric tons of oil equivalent) in 1960 to 79 mtoe in 1973. As Table 7.9 shows, this fall was far more pronounced in industries other than iron and steel. These latter industries must consume coke in the production processes. They therefore had little room to substitute oil for coal. Other industries, which used coal mainly for heating purposes, were able to substitute residual fuel oil for coal. Thus, while coal consumption in iron and steel fell only by 2 mtoe between 1960 and 1973, it declined 34 mtoe in all other industries. A steeper decline, 55 mtoe, occurred in the 'other' sector, which consisted mainly of households and office buildings.

Only in the generation of electricity did the consumption of coal rise. This rise occurred because many electrical power stations were publicly-owned. In those countries where coal production was

Table 7.8: Indigenous Solid Fuel Production and Total Energy Requirements, OECD Europe, Selected Years

Year	Production MTOE[b]	Index	Requirements[a] MTOE[b]	Index
1957	375	100	557	100
1960	334	89	622	112
1965	318	85	795	143
1970	250	67	1,032	185
1973	222	59	1,178	211
1974	214	57	1,164	209
1975	221	59	1,117	200
1976	218	58	1,182	212
1977	213	57	1,196	225
1978	212	57	1,227	220

Notes: a. Excludes consumption for overseas marine bunkers. b. Millions of metric tons of oil equivalent.
Sources: 1957: OECD, *Energy Policy: Problems and Objectives* (OECD, Paris, 1966), p.31; 1960-73: OECD, *Energy Balances of OECD Countries, 1960/1974* (OECD, Paris, 1976), pp.41, 46, 51, 54; 1974-8: IEA, *Energy Balances of OECD Countries, 1974/1978* (IEA, Paris, 1980), pp.7-11.

Table 7.9: Solid Fuel Consumption by Sector, OECD Europe, Selected Years (MTOE)

Year	Industry Iron and steel	Other	Electricity generation	Other[a]
1960	56	59	88	96
1965	55	47	110	92
1970	56	34	117	66
1973	54	25	120	41
1974	54	24	113	48
1975	44	21	112	41
1976	45	20	130	38
1977	41	19	133	38
1978	41	20	137	37

Note: a. Mainly home heating but also includes heating of office buildings, street lighting and agriculture.
Sources: 1960-73: OECD, *Energy Balances of OECD Countries, 1960/1974* (OECD, Paris, 1976), pp.41, 46, 51, 54; 1974-8: IEA, *Energy Balances of OECD Countries, 1974/1978* (IEA, Paris, 1980), pp.7-11.

important, these power stations, as a matter of public policy, increased their purchases of coal to compensate the industry for its loss of markets in the rest of the economy.[49] Nevertheless, coal's share of the growing electricity generation market fell from almost 86 per cent in 1960 to slightly more than 50 per cent in 1973. Oil's share rose from about one-tenth to more than one-third. The share of natural gas also increased significantly, as is shown in Table 7.10. The newer fossil-fuelled power plants became increasingly dependent on fuels other than coal.

Table 7.10: Structure of Fossil Fuel Consumption in Electricity Generation, OECD Europe, Selected Years (per cent)

Year	Solid fuels	Liquid fuels	Natural gas
1960	86	12	2
1965	78	20	2
1970	61	33	6
1973	52	37	11
1974	50	37	13
1975	52	33	15
1976	53	33	14
1977	56	30	14
1978	57	31	12

Sources: 1960-1973: OECD, *Energy Balances of OECD Countries, 1960/1974* (OECD, Paris, 1976), pp.41, 46, 51, 54; 1974-8: IEA, *Energy Balances of OECD Countries, 1974/1978* (IEA, Paris, 1980), pp.7-11.

The shares of the non-fossil-fuel energy sources, hydro-power and nuclear fuel, also declined during this period, reinforcing the dependence on imported fuels. In 1960, as can be seen in Table 7.11, these non-fossil-

Table 7.11: Structure of Fossil and Non-Fossil Fuel Energy Sources in the Generation of Electricity, OECD Europe, Selected Years (per cent)

Year	Non-fossil fuels Total	Hydro	Nuclear	Fossil fuels Total
1960	52	(52)		48
1965	46	(45)	(1)	54
1970	38	(36)	(2)	62
1973	35	(31)	(4)	65
1974	32	(26)	(6)	68
1975	35	(27)	(8)	65
1976	31	(23)	(8)	69
1977	37	(27)	(10)	63
1978	36	(26)	(10)	64

Sources: 1960-73: OECD, *Energy Balances of OECD Countries, 1960/1974* (OECD, Paris, 1976), pp.41, 46, 51, 54; 1975-8: IEA, *Energy Balances of OECD Countries, 1974/1978* (IEA, Paris, 1980), pp.7-11.

fuel energy sources accounted for more than one-half of total electric power generation. By 1973, their share had fallen to slightly more than one-third. Given the limited number of exploitable hydro sites in Europe, their share would have fallen even further if it were not for the expansion of nuclear power. Nuclear power's contribution has however been significant only since 1973, rising to one-tenth in 1978. Its rise just about kept even with the decline in hydro. Whether nuclear power would have (or, perhaps, should have) risen at a faster rate had European governments concentrated more resources on it is difficult to say. It does however seem fairly certain the Robinson Commission's lack of enthusiasm for nuclear energy together with the ready availability of cheap oil imports did not generate strong support for its rapid development. In any event, in electricity generation, as in the other sectors of the economy, the trend towards cheap, imported energy continued unabated during the 1960s and the early part of the 1970s. That most of these imports came from the politically unstable Middle East apparently did not worry the key policy-makers of the OECD (Organisation for Economic Cooperation and Development), the successor body to the OEEC.

OECD Study

In a 1966 study, the OECD emphasised the need to obtain 'adequate and reasonably secure supplies' of energy at 'reasonable energy prices'.[50] It further urged governments to adopt policies encouraging the 'reduction of the cost of energy to the community as a whole'.[51] Though admitting that energy imports 'clearly involve direct foreign exchange costs', it nevertheless asserted that 'it is not always possible to determine their net effects on the balance of payments, particularly in cases where they replace more expensive supplies from indigenous resources'.[52]

The study went on to say:

Thus, import restrictions on energy may lead to an increase in the price of energy within the country concerned which may prejudice the competitive position of its manufacturing industries, in particular of those in which energy costs are a relatively high proportion of total costs, such as the iron and steel industry. Therefore, the long-term solution to balance of payments problems would seem to lie instead either in making the indigenous industry competitive or *encouraging the transfer of labour and capital resources to more productive sectors, in particular manufactures for export.* (Italics added)[53]

Implicit in this statement is the concept that Europe reinforce its
position as an importer of raw material and an exporter of
manufactured products. It was, in effect, a continuation of the old
colonial doctrine that dictated that the role of the underdeveloped
countries as suppliers of cheap primary products remain unchanged.
In the post World War II world, where open imperialist transgressions
were no longer socially acceptable, the process of obtaining cheap
energy was to rely on the 'free market' mechanism rather than on
military seizure of the regions supplying the raw material. The OECD
advocated a policy of economic liberalism in energy. Private enterprise
was to play the key role in supplying energy and ensuring the proper
diversification of energy sources. 'In those countries of the area,' the
report stated, 'where oil supplies are mostly assured by private
companies, market considerations have been mainly responsible for
geographical diversification which has essentially been handled by
private industry.'[54]

Private companies, motivated by 'market considerations' would
presumably bring about the desirable results.

> In view of the large resources of energy available in the world as a
> whole, it would appear at first sight that most OECD countries
> could secure sufficient supplies at reasonable and stable prices by
> removing obstacles to international commerce and allowing each
> fuel to win its markets on the advantages it offered. Indeed,
> competition between the different fuels and freedom of choice
> for the consumer are among the basic elements in the energy
> policies of countries with or without indigenous resources.[55]

These 'rules of the game', which were adopted in varying degrees by
all the European OECD countries, were designed to favour oil. The
thrust towards private enterprise was directed against the coal
industry, which was nationalised in most European countries. The
stress on 'removing obstacles to international commerce',
'competition between the different fuels and freedom of choice'
could only benefit the international oil companies, who, because
of their joint-product, were able to manipulate the prices of the
individual products to achieve their competitive goals.

The OECD, of course, realised that there was a risk inherent in
relying too heavily on imported energy. Pointing to the Suez crisis
of 1956-7 as an example, it advocated certain safeguards against
supply interruptions. One was the geographical diversification of

supplies, which, as has been already mentioned, was to be accomplished by relying on the 'market considerations'. Another was the stockpiling of reserves, 'especially of oil' to ensure 'a measure of security against a temporary interruption of oil'.[56] A third was the use of 'reserve capacity' within the United States. This buffer was to come from the US Naval Reserve, which contained 'at least 165 million tons of proved reserves which are available at short notice' and from the individual States where, because of prorationing, there was 'a considerable reserve capacity of oil and gas production'.[57] It is interesting to note that the authors of this OECD study were seemingly unaware of the basic contradiction in their position. On the one hand, they were arguing for 'competition' and 'freedom of choice' for Europe while, on the other, they were looking to the United States, a country whose laws limited 'competition' and 'freedom of choice' in energy, to guarantee Europe's security of supply.

The use of US reserves as a buffer fitted in with Bullitt's old concept of keeping US oil in the ground for use 'as a strategic reserve'. This use would not only safeguard the United States and its allies against the consequences of a prolonged interruption of oil supplies from other areas but could also act as a powerful bargaining tool against the demands of the oil producing countries. Any attempt by them to cut back production to receive better terms could be countered by production from US reserves. The use of these US-owned reserves would, at the same time, enhance US control over oil-consuming areas.

The US Government and the oil companies were apparently able to convince the OECD that the US back-up was sufficient to meet any contingencies. The 1966 study therefore confidently concluded that 'far from revealing signs of an imminent shortage of energy, our analysis of supply and demand suggests that ample supplies will be available for OECD countries at reasonable costs to support economic growth up to 1980 and beyond'.[58]

Given these optimistic assurances, Europe began to rely more and more on imported oil. This reliance was reinforced by the increased use of automobiles. By the late 1960s, private ownership of passenger cars and trucks had become wide-spread in most countries of Western Europe. Transport thus followed the other sectors in the conversion to oil. Between 1960 and 1973 consumption of gasoline increased by 200 per cent and that of gas/diesel oil by 330 per cent.[59] This shift to oil continued unabated until 1973 when, as Table 7.12 shows, oil accounted for two-thirds of Europe's fossil fuel supply.

Because of this shift, the import content of European energy requirements rose from 47 per cent in 1960 to 83 per cent in 1973.

Table 7.12: Share of Imports in Total Energy, Oil and Coal Requirements, OECD Europe, Selected Years (per cent)

Year	Total	Oil	Coal
1960	47	93	14
1965	63	94	17
1970	79	97	21
1973	83	99	21
1974	81	100	24
1975	76	97	25
1976	78	97	23
1977	76	94	24
1978	75	93	24

Sources: 1960-73: OECD, *Energy Balances of OECD Countries, 1960/1974* (OECD, Paris, 1976), pp.41, 46, 54; 1974-8: IEA, *Energy Balance of OECD Countries, 1974/1978* (IEA, Paris, 1980), pp.7-11.

As can be seen in Table 7.12, the import content of oil, which was continuously above 90 per cent during all of the 1960s and most of the 1970s, rose from 93 per cent in 1960 to practically 100 per cent in 1974. In that year, nearly all the imports, 96 per cent, came from two geographical areas, the Middle East and North Africa.[60] Instead of the market mechanism achieving diversification in fuels and in supply sources, it achieved the opposite – the concentration on one fuel and in two geographical areas which have strong political and cultural links with each other.

End of Oil Surplus

Along with this shift to oil came another, and somewhat unexpected, event – the disappearance of the world oil surplus. Even more ominous was the evaporation of the United States surplus. Ironically, this evaporation was partly due to the imposition of import controls. These controls, as mentioned earlier, were ostensibly designed to strengthen national security. In reality, they had the opposite effect. First of all, they depleted US reserves at a faster rate than would have happened if imports had been free to enter. Secondly, they sharply accelerated the pace at which Europe converted to oil. Because of the limited access to the US market, all the oil companies, both the internationals and the independents, looked to Europe as an alternative market for their oil.

The conversion to oil lowered the elasticity of demand for that

product. With the abandonment of coal-burning equipment, consumers
of oil could not readily shift to other fuel sources if oil's price rose.
Accompanying this decline in the elasticity of demand was a fall in the
elasticity of supply. The closing-down of the coal mines foreclosed
competition from competing fuels. The exhaustion of reserves in other
areas, especially of those in the United States, prevented competitors
from increasing their oil production in response to an increase in price.
With this lowering of demand and supply elasticities, the time was
ripe for a significant increase in oil prices. The only question was when
and how this increase would be instituted.

Oil Price Increase

The Arab-Israeli War of 1973 provided the answer. If that war had not
taken place, some other event would have triggered the increase. This
increase was desired both by the international oil companies and the
producing countries. In this sense their interests converged. But beyond
this mutual desire for an increase, this convergence ended.

The significance of the increases of 1973-4 was that the oil producing
countries initiated them rather than the oil companies. Until the early
1970s the oil companies unilaterally determined both the price and the
income paid to the host countries. After 1973 the tables were turned.
The oil countries determined both the price and the income accruing
to the companies. Though the pricing decisions by OPEC brought
significant short-term financial gains to the oil companies – their profits
reached all-time highs – they also presented the companies with serious
long-term problems. Essentially the assumption of unilateral decision-
making by OPEC signalled a shift in the balance of power from the
companies to OPEC. It represented a loss of control. This loss of control
posed a serious threat to the long-term position of both the internationals
and the US Government.

Ironically, it was the policy, so vigorously pursued by both these
groups, of converting Europe from coal to oil that gave OPEC the power
it enjoys today. Had this conversion, designed to ensure US hegemony
over Europe, not taken place, OPEC could never have been able to
assume control of oil. The action of OPEC came furthermore at a time
when it was impossible for the US to act militarily against the oil-
producing countries. Bogged down in Vietnam, it was in no position
to move its forces into either the Middle East or North Africa. Because
of its policies, the control of oil slipped out of US hands. As Robert

Burns once said, 'The best laid schemes o' mice an' men gang aft a-gley'.[61]

Granted that Europe's shift to oil was policy-induced, one can still legitimately ask whether it was not also a necessary shift. Could the European economy have expanded as rapidly as it did had it been fuelled mainly by indigenous coal? Here the answer is not clear. There were certainly many hindrances to an increase in coal production, as the Robinson and 1966 OECD Reports were quick to point out. Nevertheless, there is evidence to indicate that coal could have played a much more important role than it did. The Hartley Report, in particular, emphasised the potential for coal. Odell has pointed out that in the early 1950s, the coal mines had overcome their earlier difficulties and that both production and productivity were rising. This rise was stopped in the 1950s by the inroads of oil.[62] If, perhaps, coal had been given sufficient protection, it would have been modernised and been able to compete more effectively against oil. Coal, however, was never given that chance. It was deliberately downgraded and Europe became dependent on imported oil.

It is this dependence that is at the root of the world's present-day energy crisis. It is a crisis of control. This crisis has altered world power relationships. Above all, it has threatened the dominant position of the United States in the world of oil.

Notes

1. C. Solberg, *Oil Power* (Mason/Charter, New York, 1976), pp.176-7.
2. Ibid., p.177.
3. The solid fuels include coal, lignite and peat.
4. The fossil fuels are the hydrocarbons, i.e., solid fuels, oil and natural gas. They include neither hydro-electricity nor nuclear power.
5. UNECE (United Nations Economic Commission for Europe), *The Price of Oil in Western Europe* (UNECE, Geneva, 1955), p.28.
6. P. Odell, *Oil and World Power, Background to the Oil Crisis*, 3rd edn (Penguin Books, Harmondsworth, 1974), p.100.
7. L. Armand, *Some Aspects of the European Energy Problem, Suggestions for Collective Action* (Organisation for European Economic Cooperation, Paris, 1955), p.35.
8. Ibid.
9. Ibid., p.33.
10. Ibid., p.32.
11. Ibid.
12. Ibid.
13. Ibid., p.44.
14. Ibid., p.36.
15. Ibid., pp.35-6.
16. Ibid., pp.42-3.

166   *World Oil: The Transformation of Europe*

17. Ibid., pp.59-61.
18. H. Hartley, *Europe's Growing Needs for Energy – How Can They Be Met?*
(Organisation for European Economic Cooperation, Commission on Energy, Paris
1956), p.25.
19. Ibid., p.31.
20. Ibid., p.26.
21. Ibid.
22. Ibid., p.73.
23. Calculated from: *Energy Balances of OECD Countries, 1974/1978*
(International Energy Agency and Organisation for Economic Cooperation and
Development, Paris, 1980), p.8.
24. H. Hartley, *Europe's Growing Needs for Energy*, p.31.
25. Ibid., p.25.
26. Ibid., p.27
27. Ibid., p.30.
28. Ibid., p.56.
29. Armand, *Some Aspects of the European Energy Problem*, p.46.
30. A. Robinson, *Towards a New Energy Pattern in Europe* (Organisation for
European Economic Cooperation, Energy Advisory Commission, Paris, 1960), p.5.
31. In addition to Professor Robinson, the original members of the Commission
were: Dr Hermann Abs, vice-chairman; Professor Arnaldo Angeline; Walker Cisler;
Dr Henning Fransen; Professor Pierre Uri; Nicholas Vlactiopoulos; Louis Armand
and Professor Paul de Groote. The latter two were subsequently appointed as
Chairman and Member respectively of Euratom (European Atomic Energy
Commission). Following their appointment, they resigned from the Energy
Advisory Commission and were replaced by Jacques Desrousseaux and Dr
Anthony Trepp. See ibid., pp.5, 9-10.
32. Ibid., p.34.
33. Ibid., p.60.
34. Ibid., p.62.
35. Ibid.
36. Ibid., p.61.
37. Ibid.
38. Ibid.
39. Ibid., pp.83-4.
40. Ibid., p.85.
41. Ibid., p.84.
42. Ibid., p.85.
43. Ibid., p.60.
44. Ibid., p.28.
45. Ibid., p.5.
46. Ibid., p.5.
47. W. Jensen, *Energy and the Economy of Nations* (G.T. Foulis, Henley-on-
Thames, 1970), pp.108-9.
48. *Energy Policy, Problems and Objectives* (Organisation for Economic
Cooperation and Development, Paris, 1966), p. 31; *Energy Balances of OECD
Countries 1960/1974* (Organisation for Economic Cooperation and Development,
Paris, 1976), pp.41-55; *Energy Balances of OECD Countries, 1974/1978*, pp.7-10.
49. Jensen, *Energy and the Economy of Nations*, p.137.
50. *Energy Policy, Problems and Objectives*, p.104.
51. Ibid.
52. Ibid.
53. Ibid., p.122.
54. Ibid., p.111.
55. Ibid., p.104.

56. Ibid., p.111.
57. Ibid., p.112.
58. Ibid., p.143.
59. Calculated from *1977 Oil Statistics* (IEA, Paris, 1977), Table D.
60. Calculated from ibid., Table A.
61. R. Burns, 'To a Mouse' in J. Bartlett, *Familiar Quotations*, 13th edn (Little, Brown and Company, Boston and Toronto, 1955), p.390.
62. Odell, *Oil and World Power*, p.104.

8 THE EROSION OF US DOMINANCE

US dominance, if we recall, reached its peak with the restoration of the Shah in Iran in 1954. Not only did US corporations control most of the Middle East oil but they also commanded key positions in the economies of most of the advanced capitalist countries.

Reasons for Erosion

This US dominance was however relatively short-lived. It began to erode during the late 1950s and the early 1960s. There were many reasons for this gradual erosion.

Socialist Countries

One was the growing economic strength of the Socialist countries. This strength enabled them both to resist US attempts to control their economies and to offer Third World countries an alternative to the US multinationals. Soviet assistance to Cuba, for example, provided that country with the means to withstand the US economic blockade.

European and Japanese Revival

A second reason was the revival of the war-shattered economies of Western Europe and Japan. Having recovered from the effects of the war by the mid-1950s, they began to compete fiercely with US companies for world markets. Their expansion, interestingly enough, was partially triggered by the US policy of providing cheap oil to these countries as a means of displacing coal. This cheap energy gave European and Japanese firms a competitive advantage over US firms, who, because of oil import quotas, had to pay higher prices for their energy inputs.

Their economies grew at a more rapid rate than that of the United States. As can be seen in Table 8.1, the US growth rate between 1960, when its major rivals began making a significant impact on the world economy, and 1973, the year of the oil crisis, was lower than that of any other important industrialised market economy with the exception of the United Kingdom. Japan's rate, for instance, was approximately 2½ times greater than the US one.

Table 8.1: Average Annual Growth Rates in Real Gross National Product, 1960-73, Major Industrialised Market Economies (per cent)

Country	Rate
Japan	10.5
France	5.7
Canada	5.4
Italy	5.2
Germany	4.8
United States	4.2
United Kingdom	3.2

Source: *Economic Report of the President Transmitted to the Congress January 1981* (United States Government Printing Office, Washington, 1981), p.353.

Decline in US Export Shares. Exports played a major part in this growth surge. European and Japanese products began to capture the lion's share of the increasing volume of world trade. The US share of world manufactures exports fell from 25 per cent in 1955 to 16 per cent in 1973.[1] Given the extensive US investments in overseas manufacturing facilities, this decline is, in one sense, not surprising. To the extent that US firms began producing items abroad which they had previously exported, the US export share was bound to fall. This decline thus did not necessarily signify a deterioration in the US competitive position. During this period, however, non-US multinationals grew at a faster pace than US firms.[2] This differential growth rate was a significant factor in lowering the US share. Following their gains in foreign markets, the Europeans and Japanese then invaded the US market. As a result the US surplus on merchandise trade fell precipitously. As is shown in Table 5.5, the surplus turned into a deficit in the 1970s. This decline, as already mentioned, took place before the rise of oil prices and was due mainly to an increase in manufactured imports.

Balance-of-Payments Deficit. This fall in the trade balance further aggravated the US balance-of-payments deficit. This deficit began appearing in the late 1950s. The country's large trade surplus in that period was insufficient to finance both US military expenditures and investments abroad. The US paid for this deficit with its gold and foreign currency assets. This drain on its international reserves became so pronounced that in 1971 President Nixon broke the link between gold and the US dollar. This link, which was the cornerstone of the post-war international monetary system, set the value of gold at $35 an ounce. The United States pledged to sell its gold to all governments and central banks of countries who were members of the IMF at that rate. Under this arrangement, the dollar was considered 'as good as gold'.

The large and persistent US balance-of-payments deficits brought an ever-increasing flow of dollars into the treasuries and central banks of foreign governments. As this flow continued, many governments began to lose confidence in the dollar. They accordingly traded their dollars in for gold. The US gold stock consequently fell from a high of $25 billion in 1949 to $10 billion in 1971,[3] the year the US broke its pledge to redeem gold for dollars at $35 an ounce. This action led to a restructuring of the international monetary system from one based on fixed exchange rates to a system of flexible rates.

One important outcome of the US balance-of-payments deficits was a major redistribution of international reserves. These reserves consisted of Special Drawing Rights (SDRs), positions in the IMF Fund and holdings of 'hard' currencies as well as gold. As Table 8.2 indicates, the US controlled more than half of the world's reserves in 1949. By 1971, the US share had fallen to 10 per cent. In 1973, it had declined to 8 per cent.

Table 8.2: US Share of International Reserves, Selected Years

Year	(1) Total reserves ($ billions)	(2) US reserves ($ billions)	(3) US share (2)/(1) x 100 (%)
1949	46.1	26.0	56
1952	49.3	24.7	50
1962	62.6	17.2	28
1968	77.4	15.7	20
1969	78.3	17.0	22
1970	92.6	14.5	16
1971	130.6	13.2	10
1972	158.7	13.2	8
1973	183.7	14.4	8

Source: International Monetary Fund, *International Financial Statistics*, various issues.

The beneficiaries of this redistribution were America's major industrial competitors. By 1971, the reserves of both Germany and Japan exceeded those of the United States. Germany's were $19 billion, Japan's were $15 billion while those of the United States were $13 billion. By that year more than half the world's reserves were concentrated in the countries of Western Europe.[4] It was obvious that the United States no longer held sway over the non-communist industrialised world.

Cold War

The Cold War played an important part in this process. Ironically, US strategy in the Cold War was not only to contain the Soviet Union but also to enforce US hegemony over the rest of the world. But the Cold War brought the opposite results.

Among other things, the Cold War adversely affected the US export position. Not only did it generate inflationary pressures in the United States, thereby undermining the competitiveness of US exports, but it also slowed down the rate of technological innovations in market-oriented goods. As Solo has noted, most of the expenditures in the United States on research and development, the key to technological advancement, were on military projects. He then pointed out that:

> The increase in space and military R & D may be a diversion from and, consequently, may be at the expense of R & D oriented toward economic growth. And although the ratio of R & D expenditures to national income has increased greatly since 1953, the corresponding trend of growth-oriented R & D seems actually to have been downward . . . Not only does military and space R & D divert research power from its growth-oriented counterpart, but R & D per se can draw creative resources from alternative forms of inventive and innovative activity . . . Through the advent of R & D, those who were, or who might have been, restless, probing industrialists, innovating entrepreneurs, or inventors tinkering in their shops became, instead, engineers on project teams, heads of research divisions, scientists in laboratories, or subcontractors with the task of developing a component for a complex weapons system. Something is given to one side, but lost from the other.[5]

It is interesting to note that the two countries who turned in the best export performances during the 1960s and early 1970s, Germany and Japan, were those whose military expenditures were limited by prohibitions imposed on them after the war. Neither country has allocated nearly the proportion of its resources to the military as the United States and some of its principal allies have.

The losses in exports were compounded by the extensive expenditure to maintain facilities and bases overseas. These expenditures, as well as

those incurred in the Vietnam War, transferred international reserves
from the United States to other countries, thereby further eroding
America's international economic power base.

Implications of Erosion

Until the early 1970s the US held one important trump card against
this erosion — the control of oil. With the world becoming increasingly
dependent on oil, the US was in a position to keep recalcitrant rivals
in line. The loss of control in 1973 undermined this pillar of the
American Empire. This was the essence of the oil crisis as far as the US
was concerned. The problem was succinctly stated by Walter Levy
in the following words:

> In the past, control over the international oil companies could be
> and was used as a political instrument by their home countries in
> their relationship to importing countries, such as the United
> States apparently did during the first Suez crisis or say for oil trade
> with Cuba, and so on. This possibility, to use the control over
> foreign oil for political-strategic purposes, is disappearing fast.[6]

The New Entrants

What the US was facing was the rise of new entrants into the world
oil scene. They were not the Schumpeterian innovators who deluged
the markets with new products. They were in fact imitators, who took
over a product introduced by others. Nevertheless they were
entrepreneurs, modern-day Rockefellers, in the true sense of the
word. They had recognised a change in world market conditions,
seized the opportunities presented by that change and elevated
themselves to positions of world power. As Levy put it, the world
was witnessing 'a shift of the major center of power over international
oil from the home countries to the producing countries.'[7]

The implications of this shift were enormous. Not only did it give
the OPEC countries control over a physical resource, oil, but it also
gave them control over immense financial resources. Since the price
rises of 1973-4, an increasingly greater proportion of international
reserves has been accruing to the oil-producing countries. As Table
8.3 shows, the portion of international reserves accruing to the oil-
exporting countries increased spectacularly during the 1970s, rising
from 6 per cent at the end of 1970 to more than 20 per cent at the

Table 8.3: Oil-Exporting Countries' Shares of International Reserves, Selected Years (per cent)

Year	Oil countries	Saudi Arabia
1970	6	1
1973	8	2
1978	17	5
1980	21	6

end of 1980. The oil-exporting countries have gained control of approximately one-fifth of the non-communist world's international reserves. Saudi Arabia alone possessed the same portion of reserves at the end of 1980 that all the oil-exporting countries had at the end of 1970. Its reserves were about equal to those of such industrialised countries as the United States, Japan and Italy and exceeded those of Great Britain.

This acquisition of reserves has meant that a new bourgeoisie, a bourgeoisie whose wealth is derived from mineral resources rather than from industrial production, has entered the arena of world capitalism. As such, it poses a threat to the established wealth. As this bourgeoisie accumulates greater reserves, it increases its ability to buy out the established bourgeoisie and to compete actively against them by setting up enterprises to rival theirs.

The Threat to US Investment

In countries which because of sparse populations or of fear of social change are unable to absorb large injections of developmental expenditures, there is little choice but to invest these reserves in the developed capitalist economies. Carried to the extreme, these investments could be large enough to give the oil exporting countries control over the advanced capitalist economies. In a memorandum prepared for the 1974 Pre-Summit Meeting of Economists for the White House Conference on Inflation, Levy outlined the implications of this development.

In the year ending June 30, 1975, conservative estimates imply that OPEC countries may add between $60 and $75 billion to their surplus holdings of foreign currencies. Amounts of that order would be equal to about 75 per cent of the total book value of U.S. direct foreign investments, accumulated over many decades and owned by many hundreds of private corporations, whereas

this OPEC surplus, held by a handful of governments, will arise in
only these current twelve months. The annual amount substantially
exceeds total U.S. net foreign public and private long-term and
short-term liquid and non-liquid assets. By July, 1975 gold and
foreign exchange reserves of a handful of OPEC countries will
probably be larger than those of all major industrialized nations
combined. From then on — until such time, if ever, as these oil-
exporting economies can absorb all their oil revenues — their surplus
claims on the rest of the world will go on pyramiding.[8]

Though Levy erred in his estimates of the magnitudes — the OPEC
countries did not accumulate the massive reserves he had predicted —
he nevertheless was correct about the trends. Surplus claims would
continue to pyramid, though at a slower rate than Levy thought. This
pyramiding could eventually place the OPEC countries in a position
to control the economies of the non-communist countries. Levy
therefore argued that the industrialised countries must develop a
coordinated strategy to counter this threat.[9]

  Levy was, of course, speaking from a United States point of view.
He recognised that the United States now faced a third threat to
its hegemony. The first was that posed by the communist countries.
The second was the one flowing from the increasing economic strength
of Europe and Japan. And the third was the rise in power of the oil-
exporting countries. Probably the greatest fear of Levy and the
other influential US policy-makers was that these diverse groups
might find it to their interest to join forces against the United States.

US Reaction to Erosion

To prevent this from happening, US strategists were determined to
forge a united front of the non-communist industrialised countries
against OPEC. As early as 1969, the US, anticipating the problems
that might arise from the decline in the oil surplus under its control,
tried to persuade the members of the OECD oil committee to
formulate a common approach to energy problems. Their response
was somewhat less than enthusiastic. The US suggestion was supported
by England and Holland, the home countries of British Petroleum and
Shell.[10] In subsequent years the United States made several other
attempts to achieve a coordinated policy but with no success. In early
1973, before the advent of the oil crisis, James Akins, then Director

of the Office of Fuels and Energy of the Department of State, wrote a provocative article in *Foreign Affairs* in which he stated the US arguments for such a policy. He advocated 'the formation of an international authority' which would 'avoid cutthroat competition for available energy in terms of shortage'. He feared that 'such competition could drive prices far higher than we can presently imagine'.[11]

Consumers' Cartel

He thus advocated the establishment of a consumers' 'cartel', as a form of countervailing power to OPEC. One of the tasks of this 'cartel' would be to support the international oil companies in their negotiations with OPEC. 'The producer governments,' he stated, 'have banded together in a well functioning organization. Their immediate adversaries are only the companies – an unequal contest.'[12]

This alleged impotence of the oil companies was, of course, a hyperbole, similar to Adelman's assertion that the companies have become mere tax collectors for OPEC.[13] These companies have for years been powers in their own right. Furthermore, they have always had the backing of their home governments in foreign ventures. Levy himself has admitted that they had benefited for decades 'from the immense power of the United States.'[14] This backing was reinforced early in 1971 when the State Department persuaded the US Justice Department to exempt the oil companies from anti-trust prosecutions if they collaborated with each other in their dealings with OPEC.[15]

Equally dubious was Akins's implication that the oil companies would act in the interests of consumers and that they would not raise prices on their own. 'Many consumers already believe,' he wrote, 'that the companies have not been adequately vigorous in resisting producer demands, as they could and usually did pass on to the consumer any tax increases.'[16] He therefore urged in the interests of 'consumer solidarity' that all the governments of the major oil-importing countries, not merely the home governments of the international oil companies, give backing to these companies to enable them to resist OPEC's demands. Such backing would, of course, bolster the position of these companies and their home governments, especially the government of the United States. The last thing that United States policy-makers wanted was that the importing countries bypass the oil companies and deal directly with the governments of the producing countries.

After the price escalation of 1973-4, the danger that individual importing countries would make separate deals with the oil countries became more pronounced. It was for this reason that the United States opposed the French suggestion for a meeting between the consuming and producing nations. It insisted instead that there must be an agreement among the consuming countries before any such dialogue took place. Secretary of State Henry Kissinger stated quite categorically that 'we will not go to a consumer-producer conference without prior agreement on consumer cooperation'.[17] Without such a 'prior agreement' separate deals could be negotiated between individual consumer and producer countries. After much wrangling the French accepted the concept of a 'prior agreement' at a meeting in Martinique between Presidents Ford and Giscard d'Estaing at the end of 1975.[18]

International Energy Agency

The French however refused to go along with the US suggestion to form a consumers' 'cartel'. This 'cartel', the IEA (International Energy Agency), was formed in 1974 by the major consuming nations, France excepted, for the purpose of sharing oil supplies in the event of a cut-off directed against either the group as a whole or any of its members.[19] Not only was it designed to prevent consumers from bidding up the price in times of shortages but it was also designed to protect individual countries against a selective oil boycott. The 1973-4 boycott, it must be remembered, was directed primarily against the United States and Holland in retaliation for their support of Israel. The IEA agreement therefore gave an implicit *carte-blanche* backing to US Middle East policies by assuring the US that it would not have to bear the full burden of any adverse reaction to its policies.

Oil Sharing. Article 17 of the agreement, for instance, invokes the oil sharing provisions if:

> any Participating Country sustains or can reasonably be expected to sustain a reduction in the daily rate of its oil supplies which results in a reduction of the daily rate of its final consumption by an amount exceeding 7 per cent of the average daily rate of its final consumption during the base period . . .[20]

Once Article 17 is invoked, all the members of the IEA are required to allocate to the affected country a quantity of petroleum and products 'equal to the reduction in its oil supplies which results in a reduction of final consumption over and above' 7 per cent.[21] This compulsory allocation, furthermore, cannot be effectively vetoed if the Secretariat and the Management Committee decide that a country is eligible to receive it. It comes into force automatically unless 'the Governing Board, acting by special majority, . . . decides not to activate the emergency measures'.[22] This 'special majority' requires the consent of all but two members.[23] Thus, if as few as three of the original 16 members vote to sustain the recommendations of the Secretariat and the Management Committee, the allocation provisions come into immediate effect. This provision effectively protects any country, and the United States in particular, against a selective boycott. If the allocation procedures work according to plan, any individual country would suffer no more than a 7 per cent fall in consumption. This provision removed an important constraint on the United States in its dealings with the OPEC countries. It guaranteed the country an assured source of supply in the event of a boycott. The *Oil and Gas Journal* thus commented:

> In theory, U.S. domestic oil production is available for international allocation under the emergency program, the State Department notes. In practice, however, a spokesman says, only under the most extreme emergency situation would the U.S. ever be called upon to share any of its domestic production with the other IEA countries.[24]

The journal then quoted Julius Katz, Acting Assistant Secretary of State for Economic and Business Affairs, as saying that 'in the event of a selective embargo against the U.S. which cut back our available oil by more than 7%, we would of course expect to receive oil from the other member countries'.[25]

Article 17, while protecting the U.S. against a selective boycott, actually enhances the chance of a general boycott. Given the protection afforded to the United States, the OPEC countries could retaliate against the US only by embargoing oil to all countries who are members of the IEA. This general embargo would be the sole effective way of reducing significantly the oil supplies flowing to the US.

The fear of being forced to underwrite America's Middle East policies made several nations wary of joining the IEA. Significantly, Norway, the only European country with a surplus of oil at that time, did not sign the original agreement and joined the IEA later on an associate basis, which exempted her from the energy-sharing provisions.[26] Canada, another important producer of oil, did sign the agreement. Nevertheless, there were many questions raised in Parliament about the wisdom of adhering to it. Most important of all, France, which has pursued its own Middle East policies, refused to join.

Role of Oil Companies. Another interesting feature of the IEA is the role it assigns to the oil companies. These companies are to play an important part in carrying out any emergency allocation programme. Article 6, for instance, stipulates that the Governing Board decide 'on the procedures and modalities for the participation of oil companies' in any allocation scheme.[27] Chapter VI establishes 'a framework for consultation with oil companies'.[28] Chapter V, while requiring oil companies to provide relevant information to their home governments, who in turn will make such information available to the IEA, also protects the confidentiality of the data and proprietary information, such as 'patents, trademarks, scientific or manufacturing processes or developments, individual sales, tax returns, customer lists or geological and geophysical information, including maps'.[29] These confidentiality and proprietary provisions allow both the oil companies and their home governments considerable discretion in selecting the information to be submitted to the IEA. Finally, Articles 30 and 35 stipulate that the Standing Group on the Oil Market and the Standing Group on Emergency Questions shall, in preparing their respective reports, 'consult with oil companies to ensure that the System is compatible with industry operations'.[30]

These provisions give the oil companies enormous power. As the *Oil and Gas Journal* pointed out, 'the international oil companies will have a key role in planning and carrying out any measures to share and redirect tankers and supplies if oil imports are curtailed again . . .'[31] Their power would be exercised through the Industry Advisory Board. This Board, which originally consisted of representatives of 15 oil companies, was supposed 'to provide the Agency with a range of advice on matters related to the emergency allocation procedures and emergency reserves'.[32] This working group of companies is headed by Exxon,[33] the successor to Rockefeller's Standard Oil of New Jersey.[34] This board enables Exxon to coordinate activities with those of its supposed competitors. The anti-trust implications of these moves were not lost on the US Government. The government accordingly gave the participating companies anti-trust clearance.[35] The US Government thus gave its official blessing to a new cartel of oil companies, which is to act as a counter-weight to the cartel of oil-producing countries.

The provisions of the IEA agreement also put the oil companies in a position of forcing the other governments to back them. This backing, as was noted earlier, is one of the principal objectives of US policy. It is to be achieved through the companies' control of the flow of information. This information flow, on which the IEA must base its decisions, originates with the oil companies, who pass it on to their home governments − essentially the United States, England and Holland − who, in turn, give it to the Secretariat of the IEA. Furthermore, in their consultations with the various IEA committees, the oil companies can put their own interpretation on this information flow. Since the oil companies are not disinterested observers of the IEA decision-making process, they obviously have a stake in presenting their home governments and the IEA with their own viewpoint. Their home governments, and especially the US government, also have the opportunity of influencing IEA decisions by the control and interpretation of information flows. This process has serious implications for IEA decision-making. It has the potential of repeating the situation which led the European governments, after consultation with the oil companies, to abandon coal for oil.

The dominant position of the oil companies is further reinforced by the principles underlying the allocation mechanism. Article 9, for instance, reads:

> When allocation takes place, an objective of the Program shall be that available crude oil and products shall, insofar as possible, be shared within the refining and distributing industries as well as between refining and distributing companies in accordance with historical supply patterns.[36]

This provision protects the historical position of the international oil companies. No emergency would be permitted to challenge this position. At the same time, there is no effective provision to prevent these companies from using an emergency as an excuse for raising prices. Aside from a call for 'basing the price for allocated oil on the price conditions prevailing for comparable transactions'[37] and an article stating that 'questions relating to the price of oil allocated during an emergency shall be examined by the Standing Committee on Emergency Questions',[38] there are no references to pricing policies. These policies will presumably be decided by the Standing Committee on Emergency Questions after consultation with the oil companies. Given these circumstances, one can only wonder how effective this Standing

Committee would be in protecting consumers' interests in an oil cut-off.

Floor Price. Though ostensibly formed to act as a consumers' 'cartel', the IEA apparently was not interested in forcing oil prices down too dramatically and certainly did not wish them to return to the levels prevailing in early 1973. There were both 'political' and 'economic' reasons for this reluctance.

Kissinger expressed the 'political' reasons when he warned about the effects of a sharp price decrease on two of America's most important client oil states, Saudi Arabia and Iran. In an interview held in early 1975, he expressed his fears in the following terms:

> The only chance to bring oil prices down immediately would be massive political warfare against countries like Saudi Arabia and Iran to make them risk their political stability and maybe their security if they did not cooperate. *That is too high a price to pay even for an immediate reduction in oil prices* [Italics added].
>
> If you bring about an overthrow of the existing system in Saudi Arabia and a Khadaffi takes over, or if you break Iran's image of being capable of resisting outside pressures, you're going to open up political trends that could defeat your economic objectives.[39]

Kissinger hence was willing to pay a price for political stability in the Middle East. Though unquestionably concerned about the implications of the shift in world power following the rise in oil prices, he nevertheless felt that the consequences of opposing this shift would be worse. He looked upon the IEA as a means of preventing further price increases rather than as an instrument to bring about decreases.

At the end of 1974 there was a fairly wide-spread belief that the price of oil would fall. A number of factors contributed to this belief. Among them were: (1) the steps by the IEA to decrease oil imports; (2) the recession which hit the non-communist world following the oil price increases; (3) the growing quantities of oil owned directly by the oil countries as a result of participation and nationalisation agreements; and (4) the inability of the OPEC countries to coordinate production.

Following the formation of the IEA, the participating countries agreed to reduce imports by two million barrels per day in 1975 from their 1973 levels. They also proposed further reductions of

8 and 6 per cent in 1976 and 1977 respectively from 1973.[40] This
reduction in imports exerted downward pressure on prices.

This downward pressure was intensified as a result of the fall in
economic activity. In 1974, for instance, the rate of growth in real
Gross National Product in the OECD countries fell to 0.2 per cent
from 5.9 per cent in 1973. By 1975, it was equal to −0.9 per cent,
indicating that the actual level of output had fallen.[41] As a
consequence, total energy consumption fell by 0.8 per cent in 1974
and 2.8 per cent in 1975,[42] while oil imports dropped by 3.1 per
cent and 6.7 per cent respectively in those years.[43]

These developments created a surplus of oil in the sense that more
oil could be produced than sold at the prevailing price. By the end
of 1974 the OPEC nations, according to Henry Kissinger, had shut
down one-quarter of their capacity, approximately nine billion
barrels per day, in order to eliminate this surplus.[44]

A great deal of the oil produced was moreover owned directly by
oil countries as a result of the spread of participation agreements,
under which the oil countries assumed an equity share in the
companies, and outright nationalisations. Many believed that the
OPEC countries, in their eagerness to sell this crude, would cut prices.
Levy even suggested that it might be a good idea for the oil companies
to abandon their concessions and 'reappear as competitive buyers of
crude from producing countries'. He argued that if the oil states took
possession of these concessions then 'one might expect that OPEC
unity would erode and the competing countries would eventually
compete with each other for export sales to the companies',[45] thereby
causing prices to fall.

What Levy pointed out was that the coordinating role of the
international oil companies was eroding as the role of the national oil
companies in the producing countries was expanding. As Blair has
noted, these companies have for years coordinated production as
a means of stabilising prices.[46] The erosion of this control brought
the individual oil countries into the world market as competitors.
This new role could force prices to decline.

By early 1974, the combined effects of IEA's import cutbacks,
the recession, the increasing quantities of oil in the hands of the oil
countries and the lack of coordination among these countries began
to make themselves felt. In late January, Saudi Arabia served notice
that it was going to seek a reduction in the oil price.[47] In March,
OPEC announced that it was going to freeze prices for three months,
commencing 1 April.[48] In July, Kuwait rejected all bids at an auction

for its oil because they were too low.[49] In September, Texaco and
Standard Oil of California cut production in Indonesia by 200,000
barrels a day in response to weakened market conditions.[50]

Reviewing these developments, Frank Gardner, a columnist for
the *Oil and Gas Journal*, gleefully predicted in early 1975 that oil
prices would soon come tumbling down. Gardner wrote:

> OPEC's one big objective since its inception was to shore up crude
> prices. That has now been done — overdone. No product can long
> continue to sell at a hundred times its production cost, especially
> when it's in excess supply. The competition is coming, and with it,
> lower crude prices — not at the lows of 1960, but lower.[51]

The prospect of falling oil prices threatened to undermine the IEA's
objectives of decreasing dependence on imports from the OPEC
countries. The agreement establishing the IEA called for cooperation
by the participating countries in the 'development of alternative
sources of energy such as domestic oil, coal, natural gas, nuclear energy
and hydro-electric power'.[52] The development of these sources
required investments, whose profitability would be endangered by
low OPEC oil prices. These investments had to be protected against
this contingency. There were therefore strong 'economic' reasons
for opposing a return to low oil prices.

Kissinger thus urged that consuming countries impose a floor
price on imported oil. He asked that the IEA members 'not allow
imported oil to be sold domestically at prices which would make
these new sources noncompetitive'.[53] He particularly wanted the
IEA to prevent OPEC from engaging in 'predatory price cutting'
in order to discourage investments in new energy sources.[54] His
views were echoed even more strongly by Thomas Enders, then US
representative to the IEA and later US Ambassador to Canada, who
told an oil conference in Colorado in December, 1975 that:

> Consuming nations can act together to make it more probable
> that the bulk of available energy opportunities are in fact
> exploited, by protecting them against the downside risk of
> predatory pricing by OPEC or *a return to competitive conditions
> in the international oil market*. [Italics added] [55]

This aversion to competition stands in marked contrast to the US
attitude before the escalation of oil prices. As pointed out in chapter 7,

the US, in its eagerness to have Europe convert from coal to oil, continually stressed the desirability of freedom of choice for consumers and competition in determining the forms of energy use. Such policies then coincided with the interestes of the international oil companies. In the post-escalation era, the US cautioned against the 'return to competitive conditions'. This position again coincided with the interests of these companies. Competition apparently was desirable when it benefited these companies and undesirable when it did not.

In any event the US vigorously sought to have the other IEA members agree to a floor price. When it first broached the matter, it was rebuffed. The other members, who were interested in an immediate reduction in high energy costs, refused to go along. Japan, a country with few indigenous energy resources, was most vehemently opposed to the concept. The IEA thus rejected the floor price at its meeting of March 1975.[56]

The United States nevertheless persisted in pressing for such a price. Finally, in December 1975, the IEA adopted a floor price of $7 per barrel, but not without misgivings from some of its members. There was particular objection to the relatively high floor price, which would saddle the consuming countries with heavy energy costs. As the *Oil and Gas Journal* reported:

> Some see it as an attempt to divert investment into non-OPEC energy sources. Others see it as an endorsement of OPEC action of October 1973 when prices were doubled. A $7 floor price is almost three times the price of October 1973.[57]

The floor price was part of the 'Long-Term Cooperation Programme', which stipulated that 'the Participating countries shall, as a general measure of co-operation ensure that imported oil is not sold in their domestic markets below a price corresponding to US $7/bbl'.[58] This price, which was called the MSP (Minimum Safeguard Price), could be changed by unanimous vote of the Governing Board.[59] Presumably, it could be raised if it were insufficient to ensure the profitability of investments in alternative energy sources.

The IEA members were supposed to take specific measures if the price of imports fell below the MSP. These included the imposition of specific and variable levies, import quotas and consumption taxes.[60] These provisions served as a disincentive for OPEC to make 'legitimate' price cuts, i.e. a price reduction in response to change in supply-demand relationships. If these cuts were taxed by the oil importing countries,

the OPEC countries had little chance of increasing sales through price
reductions. These reductions would merely transfer income from the
OPEC to the consuming countries. OPEC, rather than the oil companies,
would bear the full brunt of any reduction in price.

Investment Incentive. By imposing a price floor, the IEA transformed
itself from a consumers' 'cartel' to a vehicle for protecting the interests of
the energy companies. The consumers' interests in lower prices were
made subservient to the interests of these companies. This transformation
was evident in the importance attached to stimuli to investment in the
'Long-Term Co-operation Programme'. Chapter III, for instance, states
that 'the Participating Countries will, as appropriate and necessary, create
a climate favourable for investment in energy'.[61] A 'favourable climate'
is, of course, one that is highly profitable. Perhaps more important than
protection for investment in general is the *de facto* protection it gives
to investment by American companies in other IEA countries.

That this was one of the purposes of the 'Long-Term Co-operation
Programme' was made clear by Enders, who stated:

> Each of the consuming countries faces an enormous task in mobilizing
> the capital required to diminish its dependence on imported energy . . .
> But consuming countries can assist each other by making it possible
> for countries that are capital-rich but poor in energy resources to
> invest in large-scale projects in other consuming countries.[62]

The US was 'capital-rich' but 'energy poor' in the sense that it was a
net importer of energy. Its companies would certainly look forward
to investing in countries which do not have abundant capital but
are relatively rich in energy resources, like Canada, Norway and
the United Kingdom.

Many provisions of the 'Long-Term Co-operation Programme'
facilitate this movement of US companies into other IEA countries.
One stipulates that the host governments 'accord national treatment
and most-favoured-nation treatment' to foreign enterprises engaged
in energy projects.[63] Another calls on the host countries to 'avoid
the introduction of new limitations upon the extent to which such
enterprises are accorded national treatment and most-favoured-nation
treatment'.[64] These provisions are merely an institutionalisation of
the 'open door' doctrine which the US used in its disputes with
England and Holland in the 1920s. This doctrine is reinforced by
another provision calling upon the host governments to consider

'affording to nationals and companies of other Participating Countries which become parties to such a project incentives similar to those granted to its own nationals of [*sic*] companies'.[65] This provision constrains a country, like Canada, which wants to reduce the foreign control of its economy, from taking steps to encourage its own nationals and companies in energy projects.

Canada had already taken steps to reduce the inflow of foreign investment. In 1974 it established FIRA (Foreign Investment Review Agency) to decide 'whether there is or will be significant benefit to Canada in proposals by non-Canadians regarding acquisition of control of Canadian business enterprises or the establishment of new businesses in Canada'.[66] FIRA has the right to veto any foreign investment which it deems contrary to Canada's interests.

As expected, Canada objected to Chapter V of the 'Programme', which calls on IEA members to 'work towards the identification and removal of legislative and administrative measures which impair the achievement of the overall objectives of the Programme'.[67] Because of its objections, Canada was exempted from Chapter V. But its provisions apply to all other signatories.

This 'open door' policy towards investment originating in IEA member countries stands in marked contrast to the 'closed door' treatment of investments from other areas, such as the OPEC or the communist countries. The non-discriminatory provisions do not apply to them. As in the 1920s, they are used primarily to enhance the position of US companies.

Another interesting feature of the 'Programme' is its treatment of taxes. It calls on the 'host governments to take account in considering alterations of taxation and production policy as they relate to those projects of the effect of such alterations on the economics of such projects including projects already underway'.[68] This is an up-dated version of the 'minimum duty' doctrine which the United States used against Mexico in the post-World War I years. Under it, no country should apply a tax to these particular projects that it applies to the rest of the economy if this tax affects adversely the profitability of the projects. This 'minimum duty' doctrine is thus one which discriminates in favour of foreigners. As in the 1920s the US uses this doctrine or its opposite — the supposedly non-discriminatory 'open door' one — whenever it suits its needs.

Probably the most revealing provision of the 'Programme' is the one relating to the exports of energy resources from the host countries. This provision requires that the host country consider:

Guaranteeing to enterprises or other parties from other Participating Countries, through legislative and administrative actions if necessary and appropriate, the right to export to their own countries a portion of the product of the project corresponding to their participation in the project or an equitable portion as may be agreed.[69]

This paragraph along with those encouraging the international flow of investments among the IEA countries, fits in with Enders' aim of enabling the 'capital-rich' but 'energy poor' nations of the IEA to gain access to the other members' resources through investment. The US is, in effect, using the IEA as means of both halting the general erosion of its international position and of regaining control over energy resources.

Notes

1. *National Institute Economic Review*, no. 12 (1960) p.63 and no. 70 (1974), p.93.

2. S. Hymer and R. Rowthorn, 'Multinational Corporations and International Oligopoly: The Non-American Challenge' in C.P. Kindleberger (ed.), *The International Corporation* (The M.I.T. Press, Cambridge, Massachusetts, 1970), pp.57-91.

3. *Economic Report of the President Transmitted to the Congress February 1974* (US Government Printing Office, Washington, 1974), p.356.

4. Ibid., p.355.

5. Robert Solo, 'Gearing Military R & D to Economic Growth' in J. L. Clayton (ed.), *The Economic Impact of the Cold War, Sources and Readings* (Harcourt, Brace & World, Inc., New York, 1970), pp.157-64.

6. W. J. Levy, 'An Atlantic-Japanese Energy Policy', *Foreign Policy*, no. 11 (1973), p.176.

7. Ibid., p.171.

8. W. Levy, 'The Impact of Exploding Oil Costs on the World Financial and Economic System', based on a statement prepared for the Pre-Summit Meeting of Economists for the White House Conference on Inflation, September 1974, p.4.

9. Ibid., p.12.

10. J. E. Akins, 'The Oil Crisis: This Time the Wolf is Here', *Foreign Affairs*, vol. 51, no. 3 (1973), p.486.

11. Ibid., p.487.

12. Ibid., p.488.

13. M. A. Adelman, *The World Petroleum Market* (The Johns Hopkins University Press, Baltimore, 1972), p.250.

14. Levy, 'An Atlantic-Japanese Energy Policy', pp.161-2.

15. Akins, 'The Oil Crisis: This Time the Wolf is Here', p.473.

16. Ibid., p.488.

17. 'Kissinger on Oil, Food, and Trade', *Business Week*, 13 January 1975, p.67.

18. D. Evans, *The Politics of Energy, The Emergence of the Superstate* (Macmillan of Canada, Toronto, 1976), p.51.

19. OECD, 'Establishment of an International Energy Agency in the OECD', Press Release A(74) 48, Paris, 15 November 1974. The original members of the IEA were Austria, Belgium, Canada, Denmark, West Germany, Ireland, Italy, Japan, Luxembourg, the Netherlands, Spain, Sweden, Switzerland, Turkey, the United Kingdom and the United States. Of the OECD members, France and Norway were the significant non-members of the IEA. Norway, as a potentially large oil producer and exporter, had doubts about joining an organisation representing consumers' interests. Norway later joined under special arrangements. France has never joined.

20. *Agreement on an International Energy Program* (International Energy Agency, Paris, 1975), p.15.

21. Ibid., p.12.

22. Ibid., p.16.

23. Ibid., p.34.

24. 'Antitrust Shield Mulled for IEP Efforts', *Oil and Gas Journal*, vol. 72, no. 52 (1974), p.98.

25. Ibid.

26. O. Noreng, 'Friends or Fellow Travellers? The relationship of Non-OPEC Exporters with OPEC', *The Journal of Energy and Development*, vol.4, no. 20 (1979), p.327.

27. *Agreement on International Energy Program*, p.11.

28. Ibid., p.23.

29. Ibid., p.20.

30. Ibid., pp.21, 23.

31. 'Antitrust Shield Mulled for IEP Efforts', p.97.

32. 'The International Energy Agency, 18 November 1974 – 18 November 1975', OECD/IEA Press Release, Paris, no date, p.3. In subsequent years the Board's membership was expanded to approximately 30 companies.

33. 'Antitrust Shield Mulled for IEP Efforts', p.97.

34. Standard Oil (NJ) changed its name to Exxon in November 1972. See *Wall Street Journal* 25 October 1972.

35. 'Antitrust Shield Mulled for IEP Efforts', p.97.

36. *Agreement on International Energy Program*, p.13.

37. Ibid.

38. Ibid.

39. 'Kissinger on Oil, Food and Trade', p.67.

40. 'The International Energy Agency, 18 November 1974 – 18 November 1975', p.1.

41. *Economic Report of the President Transmitted to the Congress January 1978* (US Government Printing Office, Washington, 1978), p.380.

42. *Energy Balances of OECD Countries, 1974/1978* (International Energy Agency and Organisation for Economic Cooperation and Development, Paris, 1980), p.149.

43. Ibid., p.150.

44. 'OPEC Price-Cutting Protection Sought', *Oil and Gas Journal*, vol. 73, no. 6 (1975), p.40.

45. Levy, 'An Atlantic-Japanese Energy Policy', p.188.

46. J. M. Blair, *The Control of Oil* (Pantheon Books, New York, (1976), pp.98-120.

47. *National Institute Economic Review*, no. 68 (1974), p.87.

48. Ibid., p.88.

49. Ibid., no. 70 (1974), p.80.

50. Ibid., p.82.

51. F.J. Gardner, 'Watching the World', *Oil and Gas Journal*, vol. 73, no. 2 (1975), p.25.

52. *Agreement on International Energy Program*, p.25.

53. 'OPEC Price-Cutting Protection Sought', p.40.

54. Ibid.

55. 'IEA Near Choice of $6-8 as Floor Price', *Oil and Gas Journal*, vol. 73, no. 48 (1975), p.79.

56. 'IEA Rejects Specific-Oil Price Floor', *Oil and Gas Journal*, vol. 73, no. 11 (1975), p.82.

57. 'IEA Adopts Safeguard Oil Price of $7/bbl for Oil Imports', *Oil and Gas Journal*, vol. 73, no. 52 (1975), p.82.

58. *Long-Term Energy Co-operation Programme* (International Energy Agency, Paris, 1976), p.10.

59. Ibid., p.11.

60. Ibid., p.19.

61. Ibid., p.7.

62. 'IEA Near Choice of $6-8 as Floor Price', p.78.

63. *Long-Term Energy Co-operation Programme*, p.9.

64. Ibid.

65. Ibid., p.10.

66. Statistics Canada, *Canada Year Book 1978-79* (Minister of Supply and Services Canada, Ottawa, 1978), p.898.

67. *Long-Term Energy Co-operation Programme*, p.14.

68. Ibid., p.10.

69. Ibid.

9 THE DRIVE TO REGAIN DOMINANCE

The moves by the United States to restore its dominance in energy have important implications for the oil companies. The aim of the US and the IEA is to make investments in energy more profitable. This aim is bound to have a favourable impact on the balance sheets of the oil companies.

Oil Companies' Profits

Despite OPEC's success in obtaining a much higher share of oil revenues, the profits of the companies have zoomed upward. One reason is that the companies' earnings per barrel of crude increased as the posted price rose. In 1970, when the posted price of Saudi Arabian crude was $1.80, ARAMCO paid 96.3 cents in royalties and income taxes to the Saudi Government. The government take was equal to 53.5 per cent of the price. By the end of 1974, when the posted price was $11.25, the government received from royalties, income taxes and the sale of its own oil to ARAMCO approximately $10.289, equivalent to 91 per cent of the price. The per barrel earnings of ARAMCO nevertheless increased from 74 cents to 90 cents or approximately 22 per cent. These data are shown in Table 9.1.

Since most of ARAMCO's oil went to the affiliates of its owners, the posted price represented merely a bookkeeping transaction which allocated a company's total profits and costs among its various divisions. The increase in product prices, which represent arm's length transactions to third parties, was the main reason for the increase in the companies' net earnings. Though most consuming countries imposed controls on product prices, they permitted the companies to pass on their increased costs to the final consumer. Product prices thus rose as the companies increased their payments to the host countries. The companies moreover raised their prices much more than their increased payments to the oil governments. Table 9.2 indicates the extent to which the increase in product prices exceeded those of crude. The changes in product prices in this table are estimated from the gross operating revenues and refined product sales of the four private owners of ARAMCO (Exxon, Standard Oil of California, Mobil and

Table 9.1: Estimated Distribution of Revenue Received from Sale of One Barrel of Saudi Arabian Oil by ARAMCO, 1970 and 1974

	31 August 1970 ($)	(%)	1 November 1974 ($)	(%)
Posted price	1.800	100	11.250	100
Government revenues	0.963	53.5	10.289[a]	91.5
Operating costs	0.100	5.6	0.065[b]	0.6
Company's earnings	0.737	40.9	0.896	8.0
Per cent change, company's earnings, 1970-4				21.6

Notes: a. Includes the revenue received by the government from the sale of its own crude to ARAMCO. The company paid $10.665 a barrel for this crude. The figure of $10.289 in the table is a weighted average of the royalty and income tax payments of the company on its own crude plus the company's payments for the government crude. b. This is the estimated operating costs of producing the company's crude. It does not include the operating costs of the government crude.
Source: *International Petroleum Encylopedia 1975* (The Petroleum Publishing Company, Tulsa, Oklahoma, 1975), p.7.

Table 9.2: A Comparison of Estimated Increases in Price per Barrel of Refined Products Received by the Private Owners of ARAMCO[a] and Increase in the Average Posted Price per Barrel of Saudi Arabian Light Crude, 1971-9

Year	(1) Product price increase ($)	(2) Crude price increase ($)	(3) Difference (1) − (2) ($)
1971	0.864	0.395	0.469
1972	0.462	0.274	0.188
1973	1.680	0.833	0.847
1974	10.319	7.538	2.781
1975	3.776	0.156	3.620
1976	1.981	0.464	1.517
1977	2.662	0.930	1.732
1978	2.528	0.310	2.218
1979	10.190	4.600	5.590
1971-9	34.462	15.500	18.962

Notes: a. Exxon, Standard Oil of California, Texaco and Mobil.
Sources: Product Prices, based on refinery product sales and gross operating data found in *International Petroleum Encyclopedia*, various issues; Crude Prices, M. Mauksch, P. Carrewyn and C. Redman, *Energy and Europe, EEC Energy Policy and Economy in the Context of the World Energy Crisis* (2 vols., Agence Européene d'Information, Brussels, 1975), vol. 1, p.117 and F. Al-Chalabi, *OPEC and the International Oil Industry, A Changing Structure* (Oxford University Press, Oxford, 1980), pp.85-90.

Texaco) with changes in the average posted price of Saudi Arabian light crude. As can be seen, changes in product prices were greater than changes in crude oil prices in every year since 1971, the first year that the price of Saudi Arabian light crude rose from its traditional level of $1.80 per barrel. Over this nine year period, the estimated increase in the revenues received by ARAMCO's partners came to more than $34 per barrel of refined products sold, while the increases in the payments for the crude oil acquired from Saudi Arabia, the market crude which roughly reflected the movement in world crude prices, amounted to only $15.50 per barrel. The price of products during this period thus rose by $19 per barrel more than the price of crude. In some years, such as 1974, 1975, 1978 and 1979, product prices grew much more than crude acquisition costs. The differential in favour of products ranged from $2.22 in 1978 to $5.59 in 1979. While these estimates, based on the relationship between gross operating revenues and refined product sales, are admittedly rough, they are nevertheless indicative of trends. These trends are furthermore consistent with changes in the profit position of the oil companies.

A study by Chase Manhattan Bank of the earnings of 26 leading oil companies reveals that their net income after taxes rose from $7 billion in 1970 to almost $32 billion in 1979 or by 377 per cent. Furthermore, the after-tax rate of return on invested capital of these companies rose from 10.7 per cent in the former year to 24 per cent in the latter. The average rate of return for the years 1968-72 was 10.7 per cent. In the years 1973-9, the years of significant oil price increases, the average rose to 16 per cent,[1] a level 50 per cent higher than in pre-escalation years. The companies thus increased both the absolute level of their earnings and the level relative to their invested capital. They could not have made these spectacular increases in their rates of return if they had not increased their product prices more than their crude acquisition costs.

In addition, the companies were able to increase their depreciation, depletion and amortisation write-offs considerably. As Table 9.3 shows, these write-offs increased from $7 billion in 1970 to $19 billion in 1979 or by 189 per cent. These write-offs together with their net income constitute the after-tax cash flow received by the companies. This sum is the theoretical internally generated investible capital available to the firms. As can be seen, the cash flow rose from $13 billion in 1970 to $51 billion in 1979. This latter sum, received by only 26 companies, was greater than the Gross Domestic Product of

Table 9.3: After-Tax Net Income, Write-Offs and Cash Flow, 26 Major Oil Companies,[a] 1970-9 ($ billions)

Year	(1) Net Income	(2) Write-Offs[b]	(3) Cash Flow (1) + (2)
1970	6.6	6.6	13.2
1971	7.3	7.1	14.4
1972	6.9	7.5	14.4
1973	11.7	8.3	20.0
1974	16.4	10.1	26.5
1975	11.5	11.2	22.7
1976	13.1	11.2	24.3
1977	14.4	13.0	27.4
1978	15.0	15.6	30.6
1979	31.5	19.1	50.6
Total 1970-9	134.4	109.7	244.1
Per cent change 1970-9	377	189	283

Notes: a. The 26 companies are: Amerada Hess, Ashland Oil, Atlantic Richfield, British Petroleum, Champlin Petroleum, Cities Service, Clark Oil and Refining, Compagnie Française des Pétroles, Conoco, Exxon, Getty Oil, Gulf, Louisiana Land and Exploration, Marathon Oil, Mobil Oil, Murphy Oil, Petrofina, Phillips Petroleum, Royal Dutch Shell Group, Standard Oil of California, Standard Oil (Indiana), Sun, Superior Oil, Texaco, Tosco and Union Oil. All but four of these companies, British Petroleum, Compagnie Francaise des Pétroles, Petrofina and Royal Dutch Shell, are American. b. Depreciation, depletion and amortisation. Source: P. Keenan, R. Dobias, M. O'Neil and N. Anderson, *Financial Analysis of a Group of Petroleum Companies, 1979* (Chase Manhattan Bank, New York, 1980), pp.2, 22-3.

most countries in the world.

Expansion into Other Fields

More important, the investable funds available to these companies were increasing at a time when firms in other industries were encountering financial difficulties as a result of the world-wide recession following the oil price increases. This increase in the relative profitability of the oil sector enabled firms in that sector to buy out companies in other sectors. The oil companies began to acquire an ever larger share of the ownership and control of the economies of the advanced developed countries.

Rockefeller Group

This expansion into other fields actually started long before the price rises of the early 1970s. Rockefeller started it on a personal basis in the late nineteenth century. In the 1880s he had 67 major investments in non-oil industries, such as railroads, mining and banks. These investments had a market value of approximately $23 million.[2] He bought the Mesabi iron ore deposits of Minnesota, which gave him control of 60 per cent of the nation's iron ore.[3] Later he sold his Mesabi properties to Henry Clay Frick, the coal and steel magnate. Half the payment was in preferred shares of United States Steel,[4] making him the largest single shareholder in that company.[5] Over the years he acquired a controlling interest in Consolidated Coal and Colorado Fuel and Iron and invested in such diverse companies as International Harvester and General Motors.[6]

Chase Manhattan Bank. As Rockefeller grew older, he turned over control of his holdings to his son, John D. Rockefeller Jr, and to other members of his family. The Rockefeller family, as a group, expanded their interests during the 1930s. In 1930 the Rockefeller group made one of its most important purchases by acquiring between 4 and 5 per cent of the stock of Chase Manhattan Bank, enough to give the family effective control over that bank.[7] The bank is now headed by David Rockefeller, the youngest son of John D. Jr. David has been associated with Chase since 1946.[8]

Through Chase, through the family's investments and through their trust funds, the Rockefellers have acquired interests in a host of companies. An example of the interaction of these various investment vehicles is the listing of Chase Manhattan Bank as the owner of Rockefeller Center in New York City. This Center, built in the 1930s, was one of the largest investments made by the Rockefeller family. In 1974, J. Richardson Dilworth, the senior financial advisor of Rockefeller Family and Associates, a group established to oversee the investments of the family, estimated the value of Rockefeller Center at over $98 million. According to data provided by Dilworth, the family's investment in the Center is exceeded only by its investment in Exxon. Chase Manhattan Bank owns 100 per cent of the Center's stock as a trustee for the family.[9] One can only speculate on the stock in other companies Chase Manhattan owns in a fiduciary capacity for the family.

In hearings before a US Congressional committee in 1974, Dilworth listed the securities held directly by members of the family and indirectly through two trust funds established in 1934 and 1952 by John D. Rockefeller Jr for the benefit of his children and grandchildren respectively.[10] He placed the total value of these assets at slightly over $1 billion.[11] Excluded from his list were the holdings of the Rockefeller Foundation, the Rockefeller Brothers Fund, the Rockefeller Family Fund and many charitable and educational foundations which have received shares in various companies from the Rockefeller family. It is problematical to what extent the investments of these foundations remain outside the control of the Rockefellers. Professors G. William Domhoff and Charles Schwartz, in their study of the Rockefellers, have noted that the data 'strongly suggests an overlap of activities and individuals between the family's philanthropies and its business'.[12]

Dilworth furthermore omitted the holdings of the Rockefeller and the Chase Manhattan Bank in Eastern Airlines. Laurance Rockefeller, in particular, had substantial investments in that company. The *New York Times* reported that 'while the family had by no means a controlling interest in Eastern, a long association with the airline, coupled with Laurance Rockefeller's investment and that of the family bank, Chase-Manhattan, made it highly influential in determining the airline's policies'.[13]

This *New York Times* story raised the important question of how much stock the Rockefeller's have to own to gain control of a company. Laurance Rockefeller, according to the paper, owned only 125,000 shares of Eastern's 19 million shares of common stock[14] or considerably less than one per cent of the common stock outstanding. Yet, with this relatively small portion of shares, he is alleged to be 'highly influential' in the airline's affairs.

This suggests that the actual power wielded by the combined weight of Rockefellers and the Chase Manhattan Bank is much greater than the data on ownership indicate. The Dilworth data, for instance, show that the Rockefellers own only slightly more than one per cent of Exxon,[15] thereby rendering the impression that the family no longer controls the company created by John D. Rockefeller. Yet, there is no question that the family still retains control of the company. As the *Washington Post* observed:

> The Rockefeller economic power may not exist in terms of a company-by-company analysis or in terms of outright control of particular companies. But it does exist in the minds of most Americans. Most businessmen would react quite differently if a

Rockefeller-backed enterprise went into competition with them — or if they were invited into a joint participation with such in a venture — from the way they would if it were a John Doe-backed venture.[16]

In actuality, the shareholdings in Exxon of all the Rockefeller-controlled investment vehicles, including the Chase Manhattan Bank, constituted approximately 4 per cent of the company's outstanding stock, making the Rockefeller-Chase Manhattan group the single largest stockholder in the firm.[17] This bloc, while relatively small, is nevertheless sufficient to enable the Rockefeller-Chase group to exercise effective control over a company with such a widely-dispersed stock ownership as Exxon.

IBEC. Though there can be legitimate dispute about the extent to which the Rockefellers actually control many of the firms in which they have invested, there is no question about the control of one of them, IBEC (International Basic Economy Corporation). According to Dilworth, the Rockefellers owned about 79 per cent of the outstanding shares of that company.[18] Founded in Venezuela in 1947 for the ostensible purpose of promoting 'the economic development of various parts of the world, to increase the production and availability of goods, things and services useful to the lives or livelihood of their peoples, and thus to better their standard of living', it was initially financed by Creole, Shell, the Venezuelan Government and Nelson Rockefeller along with some members of his family. The two companies and the Venezuelan Government received preferred shares while the Rockefellers received common stock, thereby acquiring direct control over the company.[19] The Venezuelan Government and the oil companies later sold out their interests after the company encountered severe financial problems.[20] Nelson Rockefeller reorganised the company and turned it into a giant multinational. He established a chain of supermarkets in Venezuela, which among other things, succeeded in driving many small Venezuelan retailers out of business. Under his leadership, IBEC acquired companies in many parts of the world, including the United States. By the end of the 1960s, it operated more than 140 subsidiaries in more than 33 countries. In addition to supermarkets, IBEC ran mutual funds, insurance companies, housing construction firms and many others.[21] 'IBEC', noted Collier and Horowitz, 'was an avatar of a new business form — the U.S. multinational with subsidiaries and markets flung

across the globe — which would be a primary fact in the economic life
of the underdeveloped world in the second half of the twentieth
century.'[22] IBEC, they also noted, instead of building up the
Venezuelan economy, 'actually helped to make the country more
dependent than before on U.S. corporations and the goods they
offered'.[23] IBEC thus turned out to be the vehicle through which
the Rockefellers used their oil earnings to gain control over important
sectors of the economies of Latin American countries.

Financial Links. Rockefeller's oil wealth was also used to acquire
control over a considerable portion of the US economy, as the partial
listing of the financial links of the Rockefeller-Chase Manhattan
Group, illustrated in Table 9.4, shows. This table, compiled from a
number of sources, indicates that the Group was the largest single
shareholder in Exxon and ARCO, two supposedly competing oil
companies, and in United, Northwestern and National airlines, three
other supposedly competing companies. The Group's other holdings
read like a 'Who's Who' of American industry. It owns securities
in at least 8 oil companies and 54 non-oil companies. Though
apparently absent from such important industries as steel, automobiles
and coal, the Group has investments in leading companies in such key
industries as computers and office equipment, chemicals,
pharmaceuticals, equipment, retail trade and the news media. The
Group furthermore has financial links with a number of supposedly
competing companies in most of these industries. Oil and airlines
have already been mentioned. Other examples are pharmaceuticals,
computers and office equipment, chemicals, electrical equipment
and the media. To the extent that these financial links involve
either outright control of these companies or extensive influence in
them, the question must be raised as to how competitive these
industries are. The Group is certainly in a theoretical position to
reduce competition among these companies by coordinating their
activities.

Directorship Links. One measure of its potential to influence these
companies is membership on their boards. In their 1974 testimony
before the Congressional committee investigating the nomination
of Nelson Rockefeller for the Vice-Presidency of the United States,
Domhoff and Schwartz revealed that nine employees of Rockefeller
Family and Associates sat on the boards of some 40 corporations,
'directing an aggregate of some 70 billion dollars in assets'.[24] Their

Table 9.4: Partial Listing of Financial Links of Rockefeller-Chase Manhattan Group to Major US Companies, 1974

Oil	*Computers and office equipment*
Exxon[a]	IBM
ARCO[a]	Texas Instruments
Standard Oil of California	Xerox
Mobil	Sperry-Rand
Standard Oil (Indiana)	Hewlett-Packard
Marathon	
Conoco	*Chemical and photography*
Texaco	Eastman Kodak
	Dow
Machinery	Allied Chemicals
Caterpillar	DuPont
Allis-Chalmers	Monsanto
	Hercules
Pharmaceutical	Polaroid
Merck	
Johnson & Johnson	*Electrical*
Upjohn	General Electric
Warner-Lambert	Motorola
	Reliance Electric
Airlines	Westinghouse
United[a]	
Northwestern[a]	*Media*
National[a]	Columbia Broadcasting System
Overseas National	American Broadcasting Company
TWA	National Broadcasting Company
Delta	*New York Times*
Braniff	*Time*
Eastern	*Time-Life*
	Newsweek
Retail trade	
IBEC[a]	*Cosmetics*
Sears-Roebuck	Avon
Kresge	Chesebrough-Pond's
Penney (J.C.)	
Safeway	*Forest and paper products*
	International Paper
Other	Weyerhauser
ALCOA	
Archer-Daniels-Midland	
AT&T	
Coca-Cola	
Corning Glass	
Dun & Bradstreet	
Long Island Lighting	
Minnesota Mining & Manufacturing	

Note: a. The Group is the largest single shareholder in this company.
Source: US House of Representatives, Committee on the Judiciary, *Hearings, Nomination of Nelson A. Rockefeller*, 93rd Congress, 2nd Session (US Government Printing Office, Washington 1974), pp.848-9; US Senate, *Nomination of Nelson A. Rockefeller, Senate Executive Report No. 34* (US Government Printing Office, Washington, 1974), pp.569-79; W. Greider and T. O'Toole, 'Rockefeller Family Holdings Touch Every Economic Sphere', *Washington Post*, 22 September 1974.

data, which do not include the interlocks between members of the
Chase board and those of other companies, were, in their words
'undoubtedly incomplete'.[25] Nevertheless their data show that the
Rockefeller influence extends to industries in which the Group has
no apparent equity interests, like transportation equipment. As can
be seen in Table 9.5, Rockefeller representatives sat on the boards
of several supposedly competing companies in such industries as
transportation equipment, airlines, chemicals, finance and retail
trade. Again the question arises as to whether these representatives used
their influence to coordinate the activities of the companies on whose
boards they sat.

The Rockefellers were also in a position to influence other
companies whose top officers sat on the board of Chase Manhattan.
Among them were executives from three oil companies — Exxon,
Royal Dutch Petroleum and Standard Oil (Indiana) — and two
companies in automatic equipment — Honeywell and Hewlett-Packard.
Those firms with officers on the Chase board are listed in Table 9.6.
It should be noted that they include firms supposedly competing
with companies shown in Tables 9.4 and 9.5.

**Table 9.5: Partial Listing of Directorship Links of Employees of
Rockefeller Family and Associates, 1974**

Transportation equipment	*Finance*
American Motors	CIT Financial
Bendix	Greenwich Savings Bank
Chrysler	Lincoln First Bank
Airline	*Retail trade*
Eastern	IBEC
Seaboard World	Macy
	S.S. Kresge
Chemicals	
International Minerals and Chemicals	*Other*
Mallindrodt Chemicals	IBM
	Howard Johnson
	New York Telephone
	Rockefeller Center

Source: US House of Representatives, Committee on the Judiciary, *Hearings,
Nomination of Nelson A. Rockefeller*, 93rd Congress, 2nd Session (US
Government Printing Office, Washington, 1974), pp.769-70.

Table 9.6: Corporations Represented on Board of Directors, Chase Manhattan Bank, 1974

Automatic controls	*Oil*
Hewlett-Packard	Exxon
Honeywell	Royal Dutch Petroleum (Shell)
	Standard Oil (Indiana)

Other
Allied Chemicals
American Smelting & Refining
AT & T
Burlington Industries
Equitable Life Assurance
Federated Department Stores
General Foods

Source: W. Greider and T. O'Toole, 'Rockefeller Holdings Touch Every Economic Sphere', *Washington Post*, 22 September 1974.

Non-Oil Energy

These tables are incomplete in that they show only the linkages of the Rockefeller-Chase Manhattan Group. They do not show the acquisitions by the firms controlled or influenced by this Group. These acquisitions have significantly strengthened the economic power of the Group over the years. A complete listing of all of them would be beyond the scope of this book. It would in fact require a book in itself. What is important, as far as this book is concerned, is a listing of the important acquisitions of some of the major oil companies, whether Rockefeller-controlled or not, as a means of showing how these companies have used their profits to extend their control over the economy.

The expansion of the oil companies started in the late 1950s and early 1960s with the acquisiton of coal and uranium reserves. They obtained these reserves either through the purchase of companies owning coal or uranium properties or by buying these properties outright. By the early 1970s this trend had become so pronounced in the United States that it caused concern among the agencies enforcing the anti-trust laws. A study, prepared by the FTC (Federal Trade Commission) expressed these fears in the following words:

Some observers have expressed concern over this trend since it has coincided with an increased flexibility of fuel use which has placed all four fuels [crude oil, natural gas, coal and uranium] in competition with each other in a large number of energy consumption sectors. Establishment of the energy companies [companies exploiting more than one energy resource] has thus raised the issue of whether this movement is serving to stifle competition by reducing the number of independent decision making units that operate within the energy sector.[26]

By the time the FTC conducted its study, five oil companies had already joined the ranks of the 20 companies holding the largest coal reserves in the United States. These five companies, as is shown in Table 9.7, controlled almost one-third of the reserves owned by the top 20. The Rockefeller-Chase Manhattan Group had shares in at least two of these companies, Conoco and Exxon. Exxon is also represented on the board of Chase Manhattan. These two companies ranked fourth and fifth among firms holding coal reserves.

Table 9.7: Oil Companies' Share of Coal Reserves Owned by 20 Largest Coal Companies in the United States, 1970

Company	Reserves (billion tons)	Rank
Conoco[a]	7.7	4
Exxon[b]	7.0	5
Occidental	3.3	7
Gulf	2.6	9
Kerr-McGee	1.5	15
Total	22.1	
Total reserves of top 20	74.7	
Per cent held by oil companies	29.6	

Notes: a. Rockefeller-Chase Manhattan Group has shares in this company.
b. Rockefeller-Chase Manhattan Group is the largest single shareholder in this company. The company is represented on the board of Chase-Manhattan Bank.
Source: US Federal Trade Commission, *Concentration Levels and Trends in the Energy Sector of the U.S. Economy* (Federal Trade Commission, Washington, 1974), p.213.

A similar situation existed in uranium. In 1971, seven companies controlled 70 per cent of US uranium reserves. Among these seven were four oil companies, Getty, Gulf, Exxon and Kerr-McGee.[27] Of the 15 top US oil companies, 10 had both coal and uranium facilities in the early 1970s, 2 had only coal facilities and 3 had only uranium facilities. Of these 15 companies, the Rockefeller-Chase Manhattan Group was a stockholder in seven and the largest single shareholder in two, Exxon and ARCO. Three of these companies, Exxon, Shell and Standard Oil (Indiana) had representatives on Chase's board. The companies and their principal subsidiaries in coal or uranium mining are listed in Table 9.8.

Table 9.8: Partial List of Major Oil Companies Owning Coal and Uranium Facilities, 1970 (Coal) and 1972 (Uranium)

	Both Coal and Uranium
Company	*Subsidiaries*
Exxon[a]	Monterey Coal Co.
	Jersey Nuclear
Texaco[b]	Texas-Zinc (50%)
Gulf	Pittsburgh & Midway Coal
	C & K Coal
Mobil[b]	—
ARCO[c]	—
Getty	KGS-JV (33%)
	Petrotomics (33%)
Sun	—
Conoco[b]	Consolidation Coal
Kerr-McGee	Petrotomics (50%)
	KGS-JV (50%)
	Kermac Nuclear Fuels
	Ambrosia Lake Uranium
	Spencer Chemical
	American Lake Uranium
	Guyan Eagle
	Lakeview Mines
Sohio	Old Ben Coal
	Coal Only
Shell[d]	—
Standard Oil of California[b]	—
	Uranium Only
Standard Oil (Indiana)[e]	—
Phillips	Holly Minerals
Cities Service	—

Notes: a. Rockefeller-Chase Manhattan Group is largest single shareholder. Represented on board of Chase Manhattan. b. Rockefeller-Chase Manhattan Group is a shareholder in company. c. Rockefeller-Chase Manhattan Group is largest shareholder. d. Represented on board of Chase Manhattan. e. Rockefeller-Chase Manhattan Group is a shareholder. Represented on board of Chase Manhattan.
Sources: US Federal Trade Commission, *Concentration Levels and Trends in the Energy Sector of the U.S. Economy* (Federal Trade Commission, Washington, 1974) pp.175, 208, 211, 213, 217-18, 233; M. Tanzer, *The Race for Resources, Continuing Struggles over Minerals and Fuels* (Monthly Review Press, New York and London, 1980), pp.116-17.

Though the oil companies had made significant inroads into coal and uranium by the early 1970s, the FTC report concluded that concentration in the latter two industries was still relatively low. Nevertheless it warned:

If petroleum companies are acquiring significant amounts of coal and uranium reserves, then in the future they may control a rising fraction of coal and uranium production. In that case, energy production concentration may rise significantly as petroleum companies become more dominant in coal and uranium production.[28]

This increased concentration, the report pointed out, could lead to 'reduced interfuel competition with a consequent rise in the overall price of fuels'. It furthermore could retard technological advance since the 'petroleum firms have no incentive to pursue those technological advances in coal and uranium which represent a threat to their oil and gas operations'.[29]

Since the report was written, the oil companies, with their treasuries bulging with increased cash flows stemming from higher energy prices, have continued their penetration into coal and uranium. They have also gone into other energy fields, like solar energy. Some, like Gulf, have dropped the world 'oil' from their title. Gulf is now officially called the Gulf Company rather than the Gulf Oil Company, indicating that it now considers itself an energy rather than an oil company.

Non-Renewable Resources

Some companies have expanded from energy into the entire non-renewable resource field. These companies, anticipating that the prices of these resources will rise as they become relatively scarce over time, are looking to them as sources of future profits.

The 1970s have witnessed a number of purchases by oil companies of non-fuel mineral companies. Standard Oil (Indiana), for instance, entered into a nickel mining venture in Australia in 1973.[30] In the same year it joined a copper mining venture in Canada with Cominco.[31] Standard Oil of California purchased a one-third interest in a nickel ore project in Colombia in 1974.[32] In 1975 it bought for $333 million 20 per cent of the common stock of AMAX, one of the world's largest mining companies.[33] AMAX is a major producer of molybdenum, iron ore, coal, nickel, copper, zinc and potash as well as one renewable resource, forestry products.[34] Apparently not satisfied with its 20 per cent share, Standard offered to buy out all the shares for the sum of $3.89 billion in early 1981. The other stockholders rejected the take-over attempt.[35]

Standard in 1977 joined Amoco Mineral, a subsidiary of Standard Oil (Indiana) in forming a joint venture, Pacific Nickel Indonesia, to explore for and exploit nickel resources in Indonesia.[36] Another company interested in non-energy minerals is Gulf. In 1973 it established Gulf Energy and Minerals, a subsidiary to develop both non-petroleum energy resources and non-energy mineral resources.[37] In 1978 it bought a 51 per cent interest in Silver King Mines' Ward Mountain district.[38]

Another company entering the non-energy minerals industry is Sohio. In 1981 Sohio paid $1.77 billion to acquire Kennecott, the leading US copper producer.[39] Kennecott, through its ownership of Peabody Coal, was the third largest owner of coal reserves in the United States in 1970.[40] This purchase thus gave Sohio extensive coal as well as copper and other mineral reserves.

In 1980 Sohio's parent, British Petroleum, made the largest financial transaction in London's history when it bought the huge British mining company, Selection Trust, for more than $1 billion. British Petroleum owns 53 per cent of Sohio's shares.[41] The British Petroleum-Sohio complex hence has access to a significant store of the world's mineral resources.

Exxon has owned its own mineral-bearing properties for many years. In 1976 it found copper and zinc on its lands in Wisconsin. This find could make it one of the top ten mineral producers in the United States.[42] Its Canadian subsidiary, Imperial Oil, has acquired a lead and zinc mine in Nova Scotia and a copper mine in British Columbia. It is also taking part in a joint venture drilling for molybdenum in British Columbia.[43] Imperial's metal activities have become so important that Imperial has created a special division, Esso Minerals Canada, to handle them.[44] Its parent, Exxon, also created a subsidiary, Exxon Minerals USA, in 1977 for the same purpose.[45] In 1980 it consolidated all its mineral activities into a new unit, Exxon Minerals Group.[46]

Exxon's acquisitions however have not been nearly as important as those of ARCO, the other oil company in which the Rockefeller-Chase Manhattan Group is the largest single shareholder. In 1977 ARCO bought out Anaconda, the ninth largest copper producer in the world,[47] for $536 million in cash.[48] The combined assets of the merged company were estimated to be $9.4 billion.[49]

General Diversification

ARCO's expansion was moreover not confined to the minerals industries. It entered other fields as well. In 1976 it surprised financial circles by buying the prestigious London newspaper, the *Observer*.[50]   The acquisition of this paper had the potential of significantly expanding the company's influence. The media, after all, can have an important impact on public opinion and government policy. Apparently ARCO found that the *Observer* was perhaps too expensive a vehicle to use for this purpose. It did not turn out to be a profitable venture and ARCO sold it in 1981.

Besides newspapers, ARCO also entered transportation. In 1978, it bought a one-third interest in R.W. Miller, an Australian coal firm with shipping interests.[51] In 1979 it acquired all the shares of Northrup, a heating and air conditioning company which also produces solar energy equipment.[52]

ARCO was not the only oil company to stray far afield. It was emulated by Gulf. In 1974 Gulf established a special unit, Gulf Merchandising, to compete in the auto after market.[53] In 1976 it formed a joint venture, American Heavy Lift, with Hansa Lines to build and operate ships. Gulf owned 75 per cent of the joint company.[54] Gulf, however, was not successful in all its attempts to buy out other firms. After attempting to purchase CNA Financial in 1973, it indefinitely postponed merger talks with that company.[55] It also failed in its endeavours to acquire Rockwell, a machinery manufacturer, in 1975.[56] Perhaps most interesting of all was its unsuccessful attempt to buy Ringling Brothers' Circus from Mattel, the toy manufacturer, in 1974.[57] Apparently Gulf had considerable difficulty in finding suitable investment outlets for its bulging profits.

Exxon was another company which moved into other industries. In 1979 it purchased Xerox's office products' division.[58] Late in 1979 it acquired control of Reliance Electric, an electrical machinery manufacturer in which the Rockefeller-Chase Manhattan Group has an interest, for $1.76 billion. Before Exxon was able to complete this purchase it had to overcome the opposition of the FTC. After a long period of negotiations the FTC finally gave its consent to the merger.[59] As it expanded into new ventures, it set up a special unit, Exxon Enterprises, to handle its non-mineral activities. In 1978 Exxon Enterprises established a firm called Data Screen, to manufacture multiplex LCDs.[60] In 1979 it created Optical Information Systems to produce semiconductors and laser components.[61] One year later it formed an office system subsidiary, Exxon Information Systems.[62]

Exxon's successes in moving into new territories have not been as notable as those of Mobil. Mobil has become the most diversified of all the major oil companies through the purchases of W. F. Hall Printing Company for $50.5 million in 1978,[63] and of Marcor, a holding company whose principal subsidiaries were Montgomery-Ward, a major US retailer, and Container Corporation of America, a large packaging firm, in 1976. The latter acquisition cost approximately $800 million.[64] As with Exxon's purchase of Reliance, the US anti-trust authorities initially opposed Mobil's attempts to buy Marcor. After protracted negotiations, they finally gave up in their efforts to block the merger.[65] Mobil has thus become a major firm in US retailing and packaging.

What we have witnessed over the past few decades is a metamorphosis of the oil companies. During the late 1950s and the 1960s they gradually began to transform themselves into energy companies through the acquisition of coal and uranium facilities. Following the rise in oil prices in the 1970s, they first became mineral companies through their purchases of properties containing copper, lead, zinc, nickel and other non-ferrous metals. They later became true conglomerates through their expansion into manufacturing.

Table 9.9 gives an indication of the scope of this expansion. It lists the acquisition of some of the major companies in industries other than oil and petrochemicals, the traditional hunting grounds of the oil companies, since 1970. Since this list is by no means complete, in that it does not contain all the acquisitions by all the companies, it must be viewed merely as the tip of the iceberg.

The significance of this list lies in the implications of the investment provisions of the agreement establishing the IEA. Given the guarantees of a floor price and investment incentives for the energy companies, one can expect that these companies will assume a greater and greater control over the economies of the member countries as long as the investment provisions are observed. These provisions, which in effect ensure the profitability of these countries, will afford them the opportunity to acquire more and more companies in a wide array of industries throughout the non-communist world. The profits from these industries will further swell the flow of funds accruing to the oil companies, thereby enhancing their ability to acquire more firms. An example of this process is the reported after-tax profits of the US oil companies in 1980. In that year, eight of the top ten profit makers in the US were oil companies. In the previous year only five

Table 9.9: Entry of Some Major Oil Companies into Fields Other than Oil and Petrochemicals since 1970 (partial listing)

ARCO[a]

Non-oil energy

Forms Cenar Association, joint venture in uranium
Joins Coalcon, joint venture to convert coal to gas and liquids
Invests 50% in Centennial Hydrocarbons, joint venture with DuPont[b]
Buys Solar Technology International
Buys oil shale holdings of Equity Oil
Joins Colony Development,[c] oil shale firm
Buys Utal coal operations and reserves of General Exploration
Buys interest in R.W. Miller, a major Australian coal concern
Buys Swisher Coal, renamed Beaver Coal

Non-energy mineral

Purchases Anaconda Copper

Other

Buys *Observer*,[d] London newspaper
Buys Northrup, airconditioning and heating firm
Buys Wisconsin Centrifugal
Acquires R.N. Parsons & Sons, Parsons Transport, Parsons Leasing and Parsons Terminal

Exxon[e]

Non-oil energy

Forms joint uranium enrichment adventure with ARCO
Joins Anvil Points shale oil extraction project
Joins Avco in joint uranium project
Forms Jersey Nuclear-Avco enterprise
Exxon Nuclear joins GE in enriched uranium venture. GE later drops out
Creates jointly with Esso Standard Sekiyu, Japan, Japan Coal Liquefaction Company to build plant in Texas
Buys ARCO's 60% share in Colony Development, oil shale firm
Buys 25% in Hail Creek Project, Australia
Acquires Yeelirri Uranium Deposit, Australia
Forms Société des Applications de l'Helioénergie with Thomas CSF of Paris, France
Invests 50% in joint venture to develop uranium deposits in Midwest Lake, Canada through its Canadian subsidiary, Imperial Oil
Imperial Oil in joint venture to develop coal in British Columbia, Canada
Imperial Oil has interests in heavy oil development, Cold Lake, Canada
Imperial Oil, partner in Syncrude, joint venture to produce oil from tar sands, Canada
Imperial Oil purchases Byron Creek Collieries, a British Columbia coal mining company

Non-energy mineral

Imperial Oil acquires Granduc copper mine in British Columbia
Imperial Oil buys lead-zinc mine in Gays River, Nova Scotia, Canada

Non-energy mineral (cont.)

Imperial Oil in joint venture to develop molybdenum in Trout Lake, British Columbia

Imperial Oil in joint venture with Japanese Sumitomo interests to develop copper, zinc, lead and silver deposits in Kutcho Creek, British Columbia

Exxon buys 87 per cent of Columbia Minera Disputada de las Condes, a Chilean copper company

Other

Buys Xerox's office products division

Exxon Enterprises forms firm, Data Screen, to manufacture multiplex LCD's

Exxon Enterprises forms Optical Information Systems, a semiconductor, laser component firm

Exxon Enterprises forms Exxon Information Services, an office information subsidiary

Gulf

Non-oil energy

Forms Gulf Energy and Minerals, a non-oil energy subsidiary

Acquires Gulf United Nuclear

Buys United Nuclear Fuels

Forms company, Combustible pour Réacteurs à Haute, to manufacture nuclear fuels in France

Forms General Atomic with Shell

Forms Solvent Refined Coal International with Ruhrkohle, a German firm, and Mitsui, a Japanese firm, to build coal liquefaction plant

Non-energy mineral

Buys 51% share in Silver King Mine's Ward Mountain mining district

Gulf Canada, Canadian subsidiary, acquires coal interests in Belcourt, British Columbia and Chip Lake, Alberta

Obtains 25 per cent partnership interest in Oklahoma Nitrogen Company

Other

Subsidiary merges with Major Realty

Forms Gulf Merchandising to compete in auto after market

Forms American Heavy lift with Hansa to build and operate ships. Gulf holds 75% interest in joint company

Purchases Kewanee Industries

Mobil[f]

Non-oil energy

Forms company with Tyco Labs to develop silicon ribbon technology for solar cells. Mobil's investment is 80%

Joins joint venture, Coalcon, to develop $237 million coal company

Buys Mt. Olive & Staunton's coal assets in Illinois for $47.5 million

Other

Buys Marcor for approximately $800 million. Marcor's chief subsidiaries are Montgomery-Ward, a major retailer, and Container Corporation of America, a large packaging company

---

Other

Buys W.F. Hall Printing for $50.5 million

Phillips
Non-oil energy

Organises a subsidiary, Phillips Coal
Joins with Sun and Sohio in a joint venture to develop shale oil in Utah
Forms joint venture, Acurex Solar Energy, to develop solar energy systems
Forms Phillips Uranium
Agrees to build geo-thermal power plant for Utah Power and Light

Shell-Royal Dutch Shell Group[g]
Non-oil energy

Forms General Atomic with Gulf
Buys Seaway Coal for $65 million
Buys 50% of Massey Coal for $680 million
Buys Crow's Nest Coal in British Columbia
Other

Buys 50.5% of Alpha Text, Canadian computerised information services concern

Sohio-BP Group
Non-oil energy

Joins with Phillips and Sun in a joint venture to develop shale oil in Utah
Engaged in uranium exploration in Canada
Purchases coal properties in British Columbia
Non-energy mineral

Purchases Kennecott Cooper for $1.77 billion
Purchases Selection Trust for $1 billion

Standard Oil of California[f]
Non-oil energy

Forms joint venture with Geothermal Resources
Buys American Nuclear for $30 million
Non-energy mineral

Acquires 1/3rd interest in nickel ore project in Colombia
Buys 20% common stock of AMAX, major mineral producer, for $333 million
Forms with Amoco, subsidiary of Standard Oil (Indiana), Pacific Nickel Indonesia

Standard Oil (Indiana)[h]
Non-oil energy

Forms Rio Blanco Oil Shale as a joint venture
Buys 21% of Solarex, manufacturer of solar energy systems
Non-energy mineral

Joins nickel venture in Australia
Joins Caribou copper mining venture with Cominco in Canada
Forms with Standard Oil of California, Pacific Nickel Indonesia
Has gold reserves in Detour Lake area, Ontario, Canada
Buys Cyprus Mines
Other

Jointly forms Analog Devices/Enterprises to lend expansion capital

---

Sun

Non-oil energy

Joins with Phillips and Sohio in a joint venture to develop shale oil in Utah
Forms subsidiary, Sunoco Energy Development, for synthetic fuels

Other

Buys Stop-N-Go Foods
Buys Data Transmission for $30 million
Acquires through joint venture, SJT, St. Johnsbury Trucking company
Acquires HP International, industrial distribution firm
Acquires Carolina Co. for cash
Acquires Kar Products
Acquires Belton Dickinson, pharmaceutical concern
Acquires Atlas Screw and Speciality
Buys Weiland Computer Group
Buys Mr. Zip

Texaco[f]

Non-oil energy

Buys Wyoming coal lands of Reynolds Metals

Other

Buys General Automatic Oil Heating

Notes: a. Rockefeller-Chase Manhattan Group single largest stockholder in company. b. ARCO pulled out of this joint venture in 1977. c. Exxon purchased ARCO's 60 per cent share of Colony Development in 1980. d. ARCO sold the *Observer* in 1981. e. Rockefeller-Chase Manhattan Group largest single stockholder in company. Company represented on board of Chase-Manhattan. f. Rockefeller-Chase Manhattan Group owns shares in company. g. Company represented on board of Chase Manhattan. h. Rockefeller-Chase Manhattan Group owns shares in company. Company represented on board of Chase Manhattan.
Sources: Company reports, newspapers and trade journals.

oil companies were in the top ten.[66] A continuation of this shift of profits toward the oil companies will increase to a marked degree the economic power of these companies and of the people who control them. The power of the Rockefeller-Chase Manhattan Group, in particular, will grow significantly. With this increase in economic power we can expect an increase in political power.

Political Links

Within the United States the political power of the oil companies grew as their economic power increased. No longer were the Rockefellers the pariahs of the Establishment, as they had been in 1911, when the US Supreme Court upheld the order dissolving the old Standard Oil

Company. The Rockefellers had become the Establishment. The oil companies, in general, and the Rockefellers, in particular, began increasingly to place their imprint on all facets of American life.

Probably they were nowhere more successful than in the field of US foreign policy. This book has already shown how the US Government after World War I pursued a policy of backing the US oil companies in their overseas ventures. It has also shown how, in Saudi Arabia, US State Department officials tended to favour the major US oil companies, like ARAMCO, over the independents. A similar pattern was exhibited in Iran. As a result, very close links have developed between the State Department and the major US oil companies in the post World War II era.

Secretaries of State

The first post-war Secretary of State, for example, was Dean Acheson, who served under President Truman. Acheson was previously associated with the law firm of Covington and Burling. One of the clients of this firm was Rockefeller's Standard Oil (NJ). The founder of this firm, Judge Covington, represented most of the major companies appearing before Congressional hearings on the depletion allowance and reputedly was the author of the original depletion allowance.[67] Acheson was one of the original architects of the Cold War.

Acheson's successor was John Foster Dulles. Dulles had been a senior partner in the law firm of Sullivan and Cromwell, the main law firm of Standard Oil (NJ).[68] Before becoming Secretary of State, Dulles was Chairman of the Board of the Rockefeller Foundation.[69] In this capacity he met frequently with David Rockefeller. At one of these meetings, they decided to name a State Department official who had previously befriended Dulles, to the post of President of the Rockefeller Foundation. This official had asked Dean Acheson to have Dulles head a mission to Japan in 1949. Acheson agreed. With the election of Eisenhower, a Republican, to the US Presidency in 1952, this official, a Democrat, lost his State Department post. He was thus in a position to accept the offer by Dulles and David Rockefeller. That official was Dean Rusk.[70]

Dean Rusk remained with the Rockefeller Foundation until the election of John F. Kennedy, a Democrat, to the Presidency, Kennedy appointed Rusk as Secretary of State. Rusk continued the Cold War policies, initiated by Acheson and strengthened by Dulles. Under Rusk the US embarked on its Vietnam adventure, the most disastrous war in its history.

One result of the war was the election of Nixon, a Republican, as President in 1968. The American electorate turned to Nixon with the hope that he would extricate the country from the Vietnam quagmire. Though the election of Nixon ended Rusk's tenure as Secretary of State, it did not bring an end to the links between the State Department and the Rockefellers. Upon assuming the presidency, Nixon appointed Henry Kissinger as his national security advisor. In that position Kissinger formulated US foreign policy. Later Nixon formalised Kissinger's role by appointing him Secretary of State.

Kissinger came to Washington from the Rockefeller Brothers Fund, where he was director of its Special Studies Project. He was also chairman of the International Advisory Committee of the Chase Manhattan Bank.[71] When Kissinger left Rockefeller's office for Washington in 1969, Nelson Rockefeller gave him a gift of $50,000 in appreciation of his services.[72] Later, when Kissinger was in Washington, he and his wife spent their vacation with Nelson Rockefeller in Dorado Beach, Puerto Rico.[73] The friendship between these two men appeared indissoluble.

Under Kissinger the United States continued its war against Vietnam, expanding it into Cambodia and Laos. The US embarked on the mass bombings of Hanoi and Haiphong, which evoked world-wide protests. Nevertheless he was unable to achieve victory. The military prowess of the Vietnamese and the political turmoil in the United States forced him to accept defeat.

Part of the political turmoil centred around the 'Watergate Affair'. Though this scandal forced the resignation of Nixon, it did not touch Kissinger. He remained as Secretary of State under Nixon's replacement, Gerald Ford. The 'Watergate Affair' also tainted Ford, who lost the Presidency to Jimmy Carter, a Democrat, in 1976. The election of Carter, a Democrat, spelled the end of Kissinger's career as Secretary of State. But it did not break the links between the Rockefellers and the State Department.

President Carter appointed another Rockefeller alumnus, Cyrus Vance, as Secretary of State. Before coming to Washington, Vance had been Chairman of the Trustees of the Rockefeller Foundation, the same post held by John Foster Dulles.[74] Vance guided US foreign policy during most of the Carter years. He resigned as Secretary of State in the latter part of the Carter regime because of a disagreement over the handling of the Iranian hostage affair. Carter's handling of this affair helped pave the way for the election of Ronald Reagan to the Presidency.

Upon becoming President, Reagan appointed General Alexander
Haig as Secretary of State. This was Haig's second stint in Washington.
He was originally brought to Washington in 1969 by Henry Kissinger
to serve as Kissinger's assistant in preparing foreign policy briefings
for President Nixon.[75] While in Washington, he seemed to be well-
disposed to the Rockefellers. In 1974, when he was White House chief-
of-staff, he tried to persuade David Rockefeller to become Secretary
of the Treasury, even though Nixon had no love for the Rockefellers.
Sensing Nixon's coolness, Rockefeller declined the offer.[76]

The appointment of Haig by Reagan thus helped to maintain the
Rockefeller link to the State Department, a link that remained
unbroken throughout the entire post World War II period under both
Republican and Democrat presidents. Such a continuous link in a
democracy, which ostensibly eschews heredity as a basis for power,
is indeed remarkable. Probably no other single family in modern
history has had such an influence on the foreign policy of a major
world power. The links between the Rockefellers and the Secretary
of State are shown in Table 9.10.

**Table 9.10: Links between US Secretaries of State and Rockefeller-
Chase Manhattan Group, Post World War II Years**

Secretary	Link
Dean Acheson	Former partner, Covington and Burling, a law firm of Standard Oil (NJ).
John Foster Dulles	Former partner, Sullivan and Cromwell, major law firm of Standard Oil (NJ); Chairman, Board of Trustees, Rockefeller Foundation.
Dean Rusk	President, Rockefeller Foundation.
Henry Kissinger	Chairman, International Advisory Committee, Chase Manhattan Bank; Director, Special Studies Project, Rockefeller Brothers Fund; received $50,000 gift from Nelson Rockefeller before going to Washington to assume governmental duties.
Cyrus Vance	Chairman, Board of Trustees, Rockefeller Foundation.
Alexander Haig	Brought to Washington in 1969 by Kissinger to serve as Kissinger's assistant.

Source: *Who's Who in America*, various editions and *New York Times*, 6 October
1964 and 17 December 1980.

Other Links

The oil companies, in general, and the Rockefellers, in particular, also had other links with the State Department. Many officials, such as under secretaries and ambassadors, had oil company connections. Table 9.11 lists only some of these officials as well as others who had participated in the foreign policy decision-making process. One of the latter was Zbigniew Brzezinski, President Carter's foreign policy advisor. Brzezinski was a director of the Trilateral Commission, a 'think tank' founded by David Rockefeller.[77] According to its official statement, the Commission 'was formed in 1973 by private citizens of Western Europe, Japan and North America to foster closer cooperation among these three regions on common problems'.[78] These 'private citizens', it should be noted, represent powerful sections of the business elites of their respective countries. It was certainly in the interests of the Rockefellers to gain their acquiescence on matters of vital concern to the family. Brzezinski presumably served this purpose when he was a director of the Commission. One can only speculate as to whether he continued to serve this purpose when he became advisor to the President.

Another high official with oil connections is George Bush, Vice-President of the United States under Ronald Reagan. Since coming to office, he has assumed a role similar to that of Brzezinski and of Kissinger in the early days of the Nixon regime. Over the objections of Secretary of State Haig, President Reagan has appointed him as a *de facto* advisor on foreign affairs. Before entering public service, Bush was the president and board chairman of Zapata Off Shore Company, an oil company which he founded. He sold his interests in Zapata upon his election to Congress in 1966.[79] Nevertheless, he maintained close connections with Texas oil interests, which heavily financed his unsuccessful bid for the Presidency in 1980.[80] In the interim between his election to Congress and his assumption of the Vice-Presidency, he served as the US Ambassador to the United Nations, Chief Liaison Officer in China and Director of the CIA (Central Intelligence Agency).[81] He was also a member of David Rockefeller's Trilateral Commission.[82]

Bush was not the only CIA director with oil company connections. Allen Dulles, who headed the Agency from 1953 to 1961, also had them. He, like his brother, John Foster Dulles, was associated with the law firm of Sullivan and Cromwell, which he joined in 1926. Upon leaving the Agency in 1961, he returned to the firm and became its first chief counsel,[83] apparently well rewarded for his service to the government.

Another Sullivan and Cromwell partner who played an important foreign policy role was Arthur Dean, who was the US Delegate to the

Table 9.11: Links of Some Key US Foreign Policy Resource Personnel to Oil Interests, Post World War II Years

Individual	Oil interest link
Winthrop Aldrich, US Ambassador to Britain	Brother-in-law of John D. Rockefeller, Jr. Head, Chase Manhattan Bank
Zbigniew Brzezinski, President Carter's Assistant for National Security Affairs	Director of Trilateral Commission founded by David Rockefeller
George Bush, US Vice-President under President Reagan. Advisor to Reagan on foreign policy US Ambassador to UN Chief US Liaison Officer to China Director, Central Intelligence Agency	Co-founder and Director, Zapata Petroleum Co. President and Chairman, Zapata Offshore Co. Member, Trilateral Commission
Emil Collado, various State Department positions, including position in Office for Fuel Development, US Executive Director, World Bank Trustee, Export-Import Bank	Executive Vice-President of Exxon. Joined Exxon after leaving State Department
Arthur Dean, US Delegate to UN Chairman, US Delegation to Law of the Sea Conference, Geneva, 1958-60 Special US Ambassador to Korea Chairman, US Delegation to Disarmament Conference, 1962	Partner, Sullivan and Cromwell, main law firm of Standard Oil (NJ) Director, El Paso Natural Gas
Allen Dulles, Director, Central Intelligence Agency	Chief Counsel, Sullivan and Cromwell
Herbert Hoover, Jr, Under Secretary of State Chief architect of Iranian Consortium Special Petroleum Advisor to John Foster Dulles	Petroleum Engineer Director, Union Oil Company
Walter J. Levy, Oil Director, European Cooperation Administration Consultant to various governments Consultant to Harriman on Iranian mission Consultant to World Bank, UN and European Economic Commission Consultant to US State Department Decorated by Shah of Iran Decorated by Secretary of State Dean Rusk	Leading oil industry consultant. Has most major oil firms as client

Table 9.11 (continued)

Individual	Oil interest link

John J. McCloy, President, World Bank
  US Military Governor and High
  High Commissioner, Germany
  Advisor to President Kennedy
	Chairman of Board, Chase
	Manhattan Bank
	Member law firm, Milbank, Tweed
	Hadley and McCloy
	Firm has defended all the major
	international oil companies against
	anti-trust actions

George McGhee, Counsellor to the
  National Security Council
  Various State Department posts
  Senior Advisor, North Atlantic
  Treaty Council
  US Ambassador to Turkey
  US Ambassador to Germany
  US Ambassador-at-Large
	Sole owner of McGhee Petroleum Co
	Director of Mobil Oil

Sources: *Who's Who in America* and *Who Was Who in America*, various issues, and newspaper and magazine clippings.

United Nations General Assembly. Dean, who was a director of El Paso Natural Gas, was also the US delegate to the UN Conference on the Law of the Sea in Geneva, the US representative in post-armistice negotiations at Panmunjon, Korea and Special US Ambassador to Korea for the years 1953-4. In addition, he was chairman of the US Delegation to the Disarmament Conference of 1962.[84]

Sullivan and Cromwell was not the only law firm with both oil company and US Government links. Another was Milbank, Tweed, Hadley and McCloy. Its most illustrious member was John J. McCloy, who has been described as the 'chairman of the American "Establishment"'.[85] McCloy had also been President of Chase Manhattan Bank. His law firm represented all the major international oil companies, the 'Seven Sisters', in their anti-trust defences against US Government prosecution. McCloy's public posts included President of the World Bank and US Military Governor and High Commissioner to Germany. He was also a highly influential advisor to President Kennedy and helped shape many of Kennedy's policies.[86]

One other person who was instrumental in shaping post World War II foreign oil policy was Walter J. Levy. Levy, whom the *New York Times* dubbed 'the dean of oil consultants',[87] was, as already noted, the oil director of the Marshall Plan. It was under his administration that Marshall Plan funds were used to find markets for American-owned oil in Europe. He was also advisor to W. Averill Harriman in Iran during the

216   *The Drive to Regain Dominance*

oil crisis there in the early 1950s. Apparently the Shah of Iran greatly appreciated Levy's advice. In 1969 the Shah showed his appreciation by awarding him a high decoration.[88]

In his capacity as a private consultant, Levy had as his client the world's major oil companies and various governments. The latter included the governments of Venezuela, Canada, Alaska, Alberta, Iran, India and the United States. He has also advised such international agencies as the World Bank, the United Nations and the European Economic Community.[89]

The advice he rendered to some of these governments may have been of questionable value. For instance, it was largely on Levy's advice that Canada adopted its NOP (National Oil Policy).[90] The NOP, it should be recalled, divided the Canadian market between domestic and foreign crude. Because of the NOP, a significant quantity of Canadian crude was exported to the United States while the important refining area of Montreal remained wholly dependent on imported crude. This National Oil Policy was in reality an 'international' oil policy in the sense that Canadian oil production became integrated with the production of the international oil companies. This policy rendered Canada, a major oil producer, vulnerable to an oil cut-off. Because the NOP blocked the extension of a pipeline from the oil producing areas of Western Canada to Montreal, Canada had difficulty in supplying Canadian oil to Montreal during the 1973 embargo. Since that crisis, Canada has abandoned Levy's NOP and has built a pipeline to Montreal.

Levy also advised India to refrain from exploring for oil, and instead to let the international oil companies perform this function.[91] This was the general advice Levy gave to all underdeveloped countries. It was most explicitly stated in his famous study for the World Bank, *The Search for Oil in Developing Countries.*[92] His whole theme was that developing countries should not devote their scarce resources to the highly risky undertaking of exploration. The international oil companies were better equipped to assume this risk. The developing countries should instead concentrate on improving health, education and the quality of life. The acceptance of Levy's advice would thus give the international oil companies control of the developing nations' oil resources.

His advice was effectively refuted by Tanzer, who had worked for Esso Standard Eastern in India. Tanzer wrote:

The question . . . of whether a government should 'go it alone' cannot be solved *a priori* but only in the light of specific possibilities and resources facing the underdeveloped country. There should be no presumption that oil exploration is beyond the means of a country nor that if the government does not undertake it private companies will. To the contrary, the presumption should be that an underdeveloped country cannot afford not to undertake an oil exploration program. This is because, viewed in the light of a social return on investment analysis, oil exploration has uniquely attractive features for economic development.[93]

Tanzer also pointed out that the question of risk has been greatly exaggerated because 'the risk can be reduced considerably by relatively inexpensive preliminary surveying efforts.'[94] He also noted that:

> Exploration is not an all-or-nothing thing, but a series of steps seeking information, which can be cut short if the initial information indicates that prospects are dim. Thus, if four wells cost $10 million, but the first two give very bad results, then you can cut your losses by ending exploration, and hence your actual risk capital will turn out to be much less than the maximum.[95]

It would thus seem that the advice that Levy gave so willingly to India and other Third World countries would prevent them from using oil as an instrument of economic development. By overexaggerating the 'risks' of exploration, he urged them to adopt a policy that would simultaneously ensure that (1) the international oil companies would gain control of their oil resources and (2) these countries would remain in a perpetual state of underdevelopment. In any event, India ignored Levy's advice, set up a national oil company, the Oil and Natural Gas Commission, which did find oil.[96]

Levy also gave advice to the United States Government, for which he received no compensation. In 1968, he served as an unpaid consultant to Eugene Rostow, Under Secretary of State for Political Affairs, Anthony Solomon, Assistant Secretary of State for Economic Affairs, Covey Oliver, Assistant Secretary of State for Inter-American Affairs and Dean Rusk, Secretary of State.[97] While Levy gave advice of questionable value to Canada, India and the Third World countries in general, it apparently was of considerable value

to the United States Government. In appreciation, Secretary of State
Dean Rusk presented him in 1968 with a plaque in recognition of his
25 years of service to the State Department. Its words read, 'In
grateful appreciation for your invaluable contribution to the welfare
of the United States'.[98]

Dean Rusk was not the only Secretary of State who placed a high
value on Levy's services. Henry Kissinger also did. Turner claims that
Levy was the principal author of a speech Kissinger gave in December
1973 advocating the creation of an international energy agency.[99]
Perhaps most important of all, Levy's services have been highly valued
by the international oil companies. *Business Week* quoted one unnamed
top executive of a major international oil company as praising Levy
for being 'good, honest, and forthright'.[100] This is unquestionably an
accurate description as far as the oil companies and the United States
Government are concerned.

Levy is probably the best example of the close links between the
oil companies and the State Department of those mentioned in
Table 9.11. Some of the others, such as Emil Collado, George McGhee
and Herbert Hoover, Jr, were discussed in earlier chapters. Another
foreign policy advisor, Winthrop Aldrich, brother-in-law of John D.
Rockefeller, Jr and Ambassador to Great Britain, was also important.
But no other individual in that table had the continuous links
with all the important echelons of the State Department under
different administrations that Levy did.

Levy's activities, along with those of the others listed in Table 9.11,
show how over the years US foreign policy has developed into an oil
foreign policy, a policy designed to protect American oil interests
throughout the world. As Levy put it in discussing the Middle East,
'What is at stake is not European oil supplies, not American oil
supplies, but American ownership of oil.'[101]

Nationalisations have put an end to American ownership both in
the Middle East and in other oil-producing areas. But these
nationalisations need not prove disastrous to US interests provided
that the United States can gain control of non-OPEC energy sources.
The achievement of this control will enable it to re-establish its
hegemony in energy.

The investment and MSP provisions of the IEA agreements are the
vehicles through which the United States will attempt to regain
dominance. They provide for the free flow of energy investments
among the member states and for a guaranteed rate of return on these
investments. Most of these investments will come from American

companies, which are controlled by a few large financial groups, the most prominent of which are the Rockefellers. This free flow of investment will enable these financial groups to gain more and more control over the economies of the major consuming countries. This control will be reinforced if these investments turn out to be highly profitable. The profits will be used to buy out companies in a wide variety of industries throughout the world.

These financial groups are not satisfied with the IEA's 'floor price' of $7 a barrel. They want a much higher guaranteed price. In addition they want government financial aid for private energy projects. A 1978 report by the Trilateral Commission, for instance, states:

> If prices remain at or near current levels, there will be little economic incentive to develop readily substitutable alternative sources or to reduce demand growth as rapidly as would be desirable under the more pessimistic longer-term forecasts. For this reason, it may be appropriate for the Trilateral countries to (1) consider various mechanisms for increasing prices gradually over the next several years in those areas of the economy where it is desirable to encourage conservation and (2) provide special incentives to industry (e.g. loan guarantees, price guarantees, etc.) for the development of alternative sources of energy.[102]

There is little doubt that, given the close ties between the US oil companies and the State Department, the US Government will exert considerable pressure on other consuming nations to adopt measures similar to those suggested by the Trilateral Commission. 'You could never implement the energy policy as a purely economic matter', Kissinger noted. 'It has', he continued, 'been a foreign policy matter from the beginning.'.[103]

Notes

1. P. Keenan, R. Dobias, M. O'Neil and N. Anderson, *Financial Analysis of a Group of Petroleum Companies* (Chase Manhattan Bank, New York, 1980), pp.22-3.
2. P. Collier and D. Horowitz, *The Rockefellers: An American Dynasty* (New American Library, New York, 1976), p.52.
3. Ibid., p.23.
4. Ibid., p.54.
5. Ibid.
6. Ibid., p.68.

7. W. Greider and T. O'Toole, 'Rockefeller Family Holdings Touch Every Economic Sphere', *Washington Post*, 22 September 1974.

8. Collier and Horowtiz, *The Rockefellers*, p.308.

9. US House of Representatives, Committee on the Judiciary, *Hearings, Nomination of Nelson A. Rockefeller*, 93 Congress, 2nd Session (US Government Printing Office, Washington, 1974), pp.847, 849.

10. Ibid., p.845.

11. Ibid., pp.847-8.

12. Ibid., p.771.

13. *New York Times*, 8 December 1974.

14. Ibid.

15. US House, *Hearings, Nomination of Nelson A. Rockefeller*, p.848.

16. *The Washington Post*, 26 September 1974.

17. Greider and O'Toole, 'Rockefeller Family holdings Touch Every Economic Sphere'.

18. US House, *Hearings, Nomination of Nelson A. Rockefeller*, p.848.

19. Collier and Horowitz, *The Rockefellers*, pp.260, 262.

20. Ibid., p.263.

21. Ibid., p.264.

22. Ibid.

23. Ibid., p.263.

24. US House, *Hearings, Nomination of Nelson A. Rockefeller*, p.769.

25. Ibid.

26. US Federal Trade Commission, *Concentration Levels and Trends in the Energy Sector of the U.S. Economy* (Federal Trade Commission, Washington, 1974), p.1.

27. Ibid., p.22.

28. Ibid., p.264.

29. Ibid., p.1.

30. *Wall Street Journal*, 5 May 1973.

31. *American Metal Market*, 8 August 1973.

32. *Wall Street Journal*, 10 October 1974.

33. *Wall Street Journal*, 2 June 1975.

34. *The Financial Post Survey of Mines and Energy Resources 1980* (Maclean-Hunter, Toronto, 1980), p.18.

35. *The Globe and Mail*, 14 March 1981.

36. *American Metal Market*, 1 November 1977.

37. *Wall Street Journal*, 8 November 1973.

38. *Wall Street Journal*, 10 February 1978.

39. *The Globe and Mail*, 14 March 1981.

40. US Federal Trade Commission, *Concentration Levels and Trends*, p.213.

41. *The Globe and Mail*, 14 March 1981.

42. *Chemical Marketing Reporter*, 24 May 1976.

43. *The Financial Post Survey*, p.213.

44. *Edmonton Journal*, 13 October 1979.

45. *Oil and Gas Journal*, 17 October 1977.

46. *Journal of Commerce*, 18 June 1980.

47. M. Tanzer, *The Race for Resources, Continuing Struggles over Minerals and Fuels* (Monthly Review Press, New York and London, 1980), p.128.

48. *Wall Street Journal*, 13 January 1977.

49. *Chemical Marketing Reporter*, 17 January 1977.

50. *Wall Street Journal*, 26 November 1976.

51. *Financial Times*, 7 April 1978.

52. *Solar Engineering Magazine*, November 1979, p.46.

53. *Home Auto*; August 1974, p.10.
54. *Journal of Commerce*, 19 November 1976.
55. *Wall Street Journal*, 30 October 1973.
56. *Wall Street Journal*, 25 August 1975.
57. *Wall Street Journal*, 20 March 1974.
58. *Datamation*, October 1977, p.5.
59. *Business Week*, 8 October 1979.
60. *Electronic News*, 30 January 1978.
61. *Computer World*, 29 January 1979.
62. *Business Week*, 28 April 1980.
63. *Publishers' Weekly*, 12 November 1978.
64. *Wall Street Journal*, 7 February 1976.
65. *New York Times*, 19 August 1976.
66. 'The Energy Rip-off', *Economic Notes*, vol. 49, no. 3 (1981), p.1.
67. C. Solberg, *Oil Power* (Mason/Charter, New York, 1976), pp.75, 183.
68. R. Engler, *The Politics of Oil* (The Macmillan Company, New York, 1961), p.310.
69. *Who Was Who in America, 1951-1960* (Marquis, Chicago, 1963), p.241.
70. P. Collier and D. Horowitz, *The Rockefellers, An American Dynasty* (New American Library, New York, 1976), p.281.
71. *Who's Who in America, 1978-79* (2 vols., Marquis, Chicago, 1979), vol. 1, p.1852.
72. *New York Times*, 6 October 1974.
73. *New York Times*, 27 December 1974.
74. *Who's Who in America, 1978-9*, vol. 2, p.3362.
75. *New York Times*, 17 December 1980.
76. Collier and Horowitz, *The Rockefellers*, pp.403-4.
77. *Trialogue*, no.25 (1980/1), pp.2, 17.
78. J. Sawhill, K. Oshima and H. Maull, *Energy: Managing the Transition* (The Trilateral Commission, New York, 1978), p.1.
79. *New York Times*, 6 November 1980.
80. *New York Times*, 10 February 1980.
81. *Who's Who in America, 1978-9*, vol. 1, p.502.
82. *New York Times*, 1 April 1980.
83. *Who Was Who in America, 1969-1973* (Marquis, Chicago, 1973), p.199.
84. *Who's Who in America, 1978-9*, vol. 1, p.827.
85. A. Sampson, *The Seven Sisters, The Great Oil Companies and the World They Shaped* (The Viking Press, New York, 1975), p.165.
86. Ibid., p.166.
87. *New York Times*, 27 July 1969.
88. *Business Week*, 25 October 1969.
89. *Business Week*, 14 April 1962; *New York Times*, 27 July 1969; *Edmonton Journal*, 15 May 1975.
90. *New York Times*, 27 July 1969.
91. *Business Week*, 14 April 1962.
92. W. Levy, *The Search for Oil in Developing Countries: A Problem of Scarce Resources and Its Implications for State and Private Enterprise* (International Bank for Reconstruction and Development, 1961).
93. M. Tanzer, *The Political Economy of Oil and the Underdeveloped Countries* (Beacon Press, Boston, 1969), pp.133-4.
94. Ibid., p.133.
95. M. Tanzer, 'Oil Exploration Strategies: Alternatives for the Third World' in P. Nore and T. Turner (eds.), *Oil and Class Struggle* (Zed Press, London, 1980), pp.89-98.

96. Ibid., pp.93-4.
97. *Oil and Gas Journal*, 1 July 1968.
98. Ibid.
99. L. Turner, *Oil Companies in the International System* (The Royal Institute of International Affairs, London, 1978), pp.179, 199.
100. *Business Week*, 14 April 1962.
101. *New York Times*, 27 July 1969.
102. Sawhill, Oshima and Maull, *Energy: Managing the Transition*, p.73.
103. 'Kissinger on Oil, Food, and Trade', p.75.

10 THE ERA OF ECONOMIC WARFARE

In its drive to regain dominance, the US has entered an era of economic warfare. This warfare is directed against the OPEC countries, the non-OPEC oil-producing nations and the rest of the industrialised world. This struggle, whatever its outcome, will be a long and costly one to all sides.

The Drive against OPEC

One of the principal US aims is to destroy or seriously weaken OPEC. The OPEC members, these Schumpeterian imitators, are the new entrants to world capitalism, threatening to topple the established entrepreneurs from the pinnacles of power. The US is determined to smash them.

It intends to do this by recreating a surplus of energy under US control. This surplus can be achieved either through the development of non-OPEC energy sources or a reduction in demand for OPEC energy or some combination of both. Kissinger has stated that 'the most important' aim of US energy policy is 'to bring in alternative sources of energy as rapidly as possible so that the combination of new discoveries of oil, new oil-producing countries, and new sources of energy create a supply situation in which it will be increasingly difficult for the cartel to operate'.[1]

The introduction of new energy sources would bring downward pressure on oil prices. OPEC's only defence would be to cut back production. But, Kissinger contended, OPEC is limited in its ability to reduce production significantly. Kissinger reasoned as follows:

Many producers are dependent on their revenues for economic development. Countries that can cut production painlessly are those that are simply piling up balances. Countries that need oil revenues for their economic development, like Algeria, Iran, and Venezuela, do not have an unlimited capacity to cut their production. If the production of these countries is cut by any significant percentage, their whole economic development plan will be in severe jeopardy. Therefore, the problem of distributing the cuts is going to become more and more severe.[2]

The severity of 'the problem of distributing the cuts' could likely
lead to the break-up of OPEC as individual producing states slash prices
to sell the extra barrel of oil. This could be the result of the
nationalisations and the consequent elimination of the production
coordinating function performed by the international oil companies.

Non-OPEC Oil Countries

The most immediate prospects for alternative energy sources lie in the
non-OPEC oil countries — Great Britain, Canada, Norway and Mexico.
All but the last two are full members of the IEA. Norway, as
mentioned earlier, is only an associate member. Mexico has not joined
the IEA. Given the rules of the IEA, Britain and Canada could have
their energy resources taken over by US companies. The rules of the
IEA conflict with attempts of member countries to give priority to
their own national companies, BNOC (British National Oil Corporation)
in England and PetroCan in Canada. Norway's Statoil and Mexico's
Pemex (Petroleos Mexicanos) also stand as barriers to US penetration
of their countries. US policy-makers do not want these countries to
develop their own national energy policies. This is especially
true of Canada and Mexico. The US has for years been pressuring these
countries to adopt a North American 'continental' energy policy.

Canada

The reaction of both Canada and Mexico to such a policy has been less
than enthusiastic. Though Canada had in effect adopted a 'continental'
policy under the NOP, it scuttled it during the 1973-4 oil crisis. In 1973
Prime Minister Trudeau formally scrapped the NOP by extending the
pipeline from Edmonton to Montreal.[3] In that same year, he further
divorced Canadian oil production from its US ties by instituting export
controls on crude and products.[4] The government also imposed an
export tax on all petroleum shipped to the United States.[5] The tax
was equal to the difference between the Canadian domestic price,
which was kept below the world price, and the theoretical price of
imported oil in Chicago.

Export Controls. In 1974 the Canadian Government announced that
it would start reducing exports to the US and phase them out
completely by 1982.[6] The reduced exports were to be used to satisfy
Canada's domestic needs, especially in the Eastern provinces, which

had been wholly dependent on imported oil. As a corollary, the government adopted a policy of 'self-reliance' in oil. 'Self-reliance', a government document stated, 'means reducing our vulnerability. It means supplying energy requirements from domestic resources to the greatest extent practicable'.[7] This document thus served notice on the United States that Canada's oil resources were not 'up for grabs', that they were to be used primarily for Canadian needs rather than to be placed in an energy pool to be shared with the United States.

Petro Can. In another move toward energy independence, the Canadian Government in 1976 established a government-owned national oil company, PetroCan.[8] Since that time it has become a major force in Canada, acquiring the Canadian subsidiaries of ARCO and Phillips Petroleum, two American companies, and Petrofina, a Belgian concern.[9] The rapid growth of PetroCan has caused considerable concern in Canada's private oil sector.

National Energy Program. That concern was heightened by the introduction in 1980 of the NEP (National Energy Program). One of the purposes of this Program was to 'Canadianise' the country's oil sector, which had long been dominated by American companies. The document outlining the Program expressed its concern over foreign control over the nation's energy resources in the following words:

Of the top 26 petroleum companies in Canada, 17 are more than 50 per cent foreign owned and controlled, and these 17 account for 72 per cent of Canadian oil and gas sales. This is a degree of foreign participation that would not be accepted — indeed, simply is not tolerated — by most other oil-producing nations.[10]

The NEP proposes to attain 'at least 50 per cent Canadian ownership of oil and gas production by 1990'. It also has as its goals the achievement of 'Canadian control of a significant number of the larger oil and gas firms' and 'an early increase in the share of the oil and gas sector owned by the Government of Canada'.[11] Under the NEP, PetroCan is to 'acquire several of the large foreign-owned firms'.[12] In addition, it will acquire a 25 per cent ownership in the new oil potential territory currently being explored in the Arctic and off the East coast.[13] To encourage private Canadian firms to play a more active role in the industry, the NEP intends to abolish the depletion allowance in most parts of Canada and substitute for it a system of incentive payments to firms engaged in exploration and development. Except in the Arctic and offshore areas, foreign-owned firms will receive no payments at all. The amounts other firms receive will be dependent on the degree of Canadian ownership. Though foreign-owned firms in the Arctic and the offshore regions will receive some incentive payments, they will receive far less than Canadian-owned ones.[14]

These payment provisions have served as an incentive for private Canadian firms to buy out control of the Canadian subsidiaries of American firms. One Canadian firm, Dome Petroleum, purchased sufficient shares of stock in Conoco to force that company to sell to it Conoco's Canadian subsidiary.[15] In mid-1981, several other Canadian firms were using similar tactics to force the American companies to sell off their Canadian subsidiaries. Not surprisingly, legislation has been introduced into the US Congress to thwart this practice. Those US oil companies demanding protection against what they term the Canadian 'buccaneering' were represented by George Ball, a former Under Secretary of State.[16]

Because these moves threatened, in Levy's words, 'the American ownership of oil' in Canada, they have aroused the concern of the United States Government. The United States has protested vigorously to the Canadian Government. According to the *Globe and Mail*, a Toronto newspaper, an unnamed State Department official complained that the NEP 'is a very large and massive retreat from the principles of national treatment'. He threatened retaliation, pointing out that there are 'many examples of significant Canadian investments in the U.S. which could be taxed, regulated or legally constrained, one way or another'.[17]

There is thus an interesting replay of history going on. The dispute with Canada over non-discriminatory treatment of US companies and over 'reciprocity' is strikingly similar to the disputes with Great Britain and Holland in the years following World War I. Regardless of how this dispute will be resolved, the United States can no longer be certain that Canada's energy resources are 'secure'. The NEP is a culmination of a trend that has been growing in Canada for the past two decades, the trend toward economic nationalism. It reflects, in part, the desire of a portion of the Canadian bourgeoisie to use Canada's resources for their own development rather than for the furtherance of US energy policy.

Mexico

Nor can the United States look to Mexico. Because Mexico nationalised
its oil in 1938, it does not have the problem of foreign ownership facing
Canada. Nevertheless the United States became interested in Mexican
oil after the large discoveries made by Pemex in the mid-1970s.
Though nationalised, Mexican oil was an alternative to OPEC oil and
was much nearer to the United States than oil from the OPEC countries.
It had the potential of being used by the US in its bargaining with
OPEC.

Despite efforts by the US and other industrialised countries for
Mexico to increase its production, the Mexican Government has
adopted a deliberate policy of limiting its output to between 2.2 and
2.5 million barrels per day.[18] It has also limited its exports to about
1.5 million barrels per day.[19] Mexico furthermore forced the US to
buy its natural gas at a relatively high price as a condition to buy
more oil.[20] Mexico has demanded that the US give its manufactured
exports special treatment in the US market and that it accept more
Mexican immigrants. The US is insisting that Mexico abide by the
rules of GATT (General Agreement on Trade and Tariffs), which
prohibit such special treatment. Mexico had refused to join GATT.[21]

There is no question that the US regards Mexican oil as a very
important part of its overall energy strategy. It is no coincidence
that President Reagan arranged to meet Mexico's President Lopez
Portillo before Reagan was inaugurated. Whether the US will
succeed in bringing Mexican oil under its control is problematical.
The Mexican bourgeoisie, like significant sections of the Canadian,
want to use their oil for their own purposes. They are bound to
demand a high price for access to their oil. The United States is
thus faced with a struggle with both its neighbours over energy
policy.

Others

This is not the only potential area of conflict confronting the US.
Another is South East Asia, a potential source of non-OPEC oil. Part
of the *rapprochement* between the United States and China involves
oil. The Chinese Government has granted a number of American
companies the right to explore for oil in the South China Sea. Among
them are: Standard Oil (Indiana), ARCO, Mobil, Exxon, Texaco,
Chevron and Phillips. Much of the exploration is being conducted
around Hainan Island, which is in territory claimed by Vietnam.[22]
The question arises as to what Vietnam will do if oil is discovered and

drilling rigs are set up. A conflict could arise between China and Vietnam over this area, which would carry a high potential for US involvement. Furthermore, given the rising nationalism in China, there is a strong possibility that, if oil is discovered, the Chinese may well demand it be used for purposes other than those desired by the US. The US may thus be involved in a struggle with China over oil.

No matter where the US may go in the non-OPEC world, whether it be Canada, Mexico, China, Vietnam, the United Kingdom, Australia, Norway or any other oil-bearing region, it will have difficulty in gaining control over world oil. Neither the United States nor its oil companies will ever be able to regain the dominance they enjoyed in the pre-OPEC days. Though the interests of all these non-OPEC oil regions differ in many respects, they have one interest in common: they do not want to see the re-emergence of an oil surplus controlled by the US and its companies. This surplus, after all, could be used against them as well as against OPEC. Assuming that the US succeeds in breaking up OPEC, then it will be in a position to play one oil country off against another. Canada's demands could then be ignored just as easily as Ecuador's. Given this community of interests, it is highly unlikely that these countries will readily accede to US dominance in oil.

Industrialised Countries

There is also a continuing struggle between the US and the other industrialised market economies as well as among these economies themselves. This conflict involves the race for new energy technologies, technologies that will eventually replace oil as the motive force of industrial societies.

Though the predictions in recent years of an imminent oil shortage have been exaggerated, there is no doubt that the supply of oil is limited and that it will eventually be exhausted. Therefore society has to move from an oil to an non-oil era. In the more distant future, society will have to look increasingly for non-fossil-fuel energy sources. The country that develops the new energy sources will be the one in the best position to assume a dominant role in the world. The major industrialised nations are therefore in a scramble to develop these new sources. The United States, to regain its dominance; the others, to gain power on their own.

This is why a coordinated energy policy is impossible. It is feasible only as a means of opposing OPEC's immediate demands. But even here unanimity has seldom, if ever, been achieved. France, for instance, has refused to join the IEA.

The war against OPEC moreover requires an energy surplus outside the control of OPEC. The building-up of that surplus through non-OPEC oil and alternative energy sources takes time. In the short term a surplus can be achieved only through a reduction in demand for energy. That reduction can come about in only two ways. The first is the more efficient use of energy, i.e. what is normally considered to be conservation.[23] The second, and really the most effective, is a general reduction in the demand for all goods and services, i.e. in a recession. The reappearance of an oil 'surplus' in 1980 and the subsequent price cuts were directly related to the recession in the developed countries.

Strength through Misery

This downturn must be viewed as part of the economic warfare against OPEC. Kissinger actually spoke of achieving 'strength through misery'.[24] Levy noted that 'a recession blamed on OPEC' might have other advantages. 'It might', he pointed out, 'gain support or acquiescence from other importing nations for American threats of economic or even military action against some Gulf Arab countries'.[25] Since Levy made that statement, the US has been building up a 'rapid deployment force' for use in the Middle East but has, so far, achieved only limited success in obtaining co-operation from the other industrialised powers.[26] But World War II, it should be recalled, arose out of the Depression of the 1930s. A severe recession in the 1980s could also lead to military confrontation.

The policy of 'strength through misery' has a purpose in addition to the one of effectuating an immediate drop in the demand for energy. It is also necessary for the development of alternative energy sources. Not only will this development require advanced technology but it will also need enormous amounts of capital. To raise these sums, national income has to be redistributed away from consumption and into investment. This is probably the main reason why the leading industrial nations have abandoned Keynesian remedies for the recession.

A Keynesian remedy implies an expansion of public-sector spending as a means of raising consumption. This public-sector spending can come only from potential investment funds. The thrust today, the

so-called 'supply side economics', is to generate more investment
funds through a reduction in public spending. It might be viewed
as a capitalist version of Stalin's policy of rapid industrialisation.
This is the rationale behind the 'anti-inflationary' and 'austerity'
programmes carried out by all the market economies since the
mid-1970s.

How long this economic warfare will last is hard to say. But as
long as it does we can expect a continuous assault on living standards
in all the industrialised countries. With this assault there will occur
growing unemployment, labour unrest, civil strife, political
instability and increased tension among the major industrialised
market powers. Whether the ultimate outcome will be the
transformation of the economic conflict into a military one is
impossible to predict.

One thing however seems clear. The end of the oil era spells
the end of the age of American dominance. The US can never
regain its lost power. The American Empire, like all the empires
of the past, will fade into oblivion. It remains to be seen whether
this Empire will end with a 'bang' or a 'whimper'.

Notes

1. 'Kissinger on Oil, Food, and Trade', *Business Week*, 13 January 1975, p.67.
2. Ibid.
3. 'Canada', *International Petroleum Encyclopedia*, 1974 (The Petroleum
Publishing Company, Tulsa, Oklahoma, 1974), pp.124-5.
4. Canada, *An Energy Strategy for Canada: Policies for Self-Reliance*
(Minister of Energy Mines and Resources, Ottawa, 1976), p.152.
5. Ibid.
6. 'Canada', *International Petroleum Encyclopedia*, 1975 (The Petroleum
Publishing Company, Tulsa, Oklahoma, 1975), p.134.
7. Canada, *An Energy Strategy for Canada*, p.124.
8. Ibid., p.155.
9. *Energy Update*, 13 February 1981.
10. Canada, *The National Energy Program* (Energy, Mines and Resources
Canada, Ottawa, 1980), p.19.
11. Ibid., p.49.
12. Ibid.
13. Ibid., p.50.
14. Ibid., pp.38-41.
15. *Energy Update*, 5 June 1981.
16. *Globe and Mail*, 22 June 1981.
17. *Globe and Mail*, 22 June 1981.
18. 'Oil Boom in Mexico May Tilt Economic Picture', *International Petroleum
Encyclopedia*, 1980 (The Penn Well Publishing Company, Tulsa, Oklahoma, 1980),
p.437.
19. *Globe and Mail*, 2 July 1981.

20. 'Oil Boom in Mexico May Tilt Economic Picture', *International Petroleum Encyclopedia*, p.438.

21. *Globe and Mail*, 29 June 1981.

22. *The Boston Globe*, 12 May 1980.

23. Strictly speaking, the more efficient use of energy is not synonymous with conservation. An increase in efficiency implies a higher output from a given input. Conservation, on the other hand, relates to the temporal allocation of a resource so as to maximise the present value of its income stream. Since the present value of income received in the distant future is low, an optimal conservation policy might dictate producing more in the present and less in the future. To the extent that an increase in efficiency might reduce present production and consumption, it might conflict with an optimal conservation policy.

24. W. J. Levy Consultants Corporation, 'Oil on New Terms', New York, 1974, p.21.

25. Ibid., p.22.

26. 'The Oil Crisis: Is There a Military Option?', *The Defense Monitor*, vol 8, no. 11 (1979), pp.1-7.

SELECT BIBLIOGRAPHY

Adelman, M.A., 'Efficiency of Resource Use in Crude Petroleum', *The Southern Economic Journal*, vol. 31 (1964)
——, *The World Petroleum Market* (The Johns Hopkins University Press, Baltimore, 1972)
Agreement on an International Energy Programme (International Energy Agency, Paris, 1975)
Akins, J.E., 'The Oil Crisis: This Time the Wolf is Here', *Foreign Affairs*, vol. 51, no. 3 (1973)
Armand, L., *Some Aspects of the European Energy Problem, Suggestions for Collective Action* (Organisation for European Economic Cooperation, Paris, 1955)
Blair, J., *The Control of Oil* (Pantheon, New York, 1976)
Canada, Royal Commission on Energy, *Second Report to His Excellency the Governor General in Council* (Ottawa, 1959)
Collier, P. and D. Horowitz, *The Rockefellers: An American Dynasty* (New American Library, New York, 1976)
Concentration Levels and Trends in the Energy Sector of the US Economy (US Federal Trade Commission, Washington, 1974)
de la Tramerye, P., *The World Struggle for Oil* (Knopf, New York, 1924)
An Energy Strategy for Canada: Policies for Self-Reliance (Minister of Energy, Mines and Resources, Ottawa, 1976)
Engler, R., *The Politics of Oil, A Study of Private Power and Democratic Directions* (The Macmillan Company, New York, 1961)
Evans, D., *The Politics of Energy, The Emergence of the Superstate* (Macmillan of Canada, Toronto, 1978)
Fanning, L., *The Shift of World Petroleum Power Away from the United States* (Gulf Oil Company, Pittsburgh, 1958)
Frank, H., 'The Pricing of Middle East Crude Oil', Ph.D. Thesis, Columbia University, 1961
Fraser, H., *Diplomatic Protection of American Petroleum Interests in Mesopotamia, Netherlands, East Indies, and Mexico* (US Government Printing Office, Washington, 1945)
Galbraith, J.K., *American Capitalism*, Sentry edn (Houghton Mifflin, Boston, 1962)
——, 'Power and the Useful Economist', *American Economic Review*,

vol. 63, no. 1 (1973)

Gols, A., 'United States Foreign Oil Investments', unpublished Ph.D. Thesis, University of Oregon, 1961

Gray, E., *Impact of Oil, The Development of Canada's Oil Resources* (The Ryerson Press/Maclean Hunter, Toronto, 1969)

Halliday, F., *Iran Dictatorship and Development* (Penguin Books, Harmondsworth, 1979)

Hartley, H., *Europe's Growing Needs for Energy – How Can They Be Met?* (Organisation for European Economic Cooperation, Paris, 1956)

Hartshorn, J.E., *Oil Companies and Governments, an Account of the International Oil Industry in Its Political Environment* (Faber and Faber, London, 1962)

Hobson, C.K., *Export of Capital* (Constable, London, 1914)

Hymer, S. and R. Rowthorn, 'Multinational Corporations and International Oligopoly: The Non-American Challenge' in C.P. Kindleberger (ed.), *The International Corporation* (The M.I.T. Press, Cambridge, Mass., 1970)

Issawi, C. and M. Yeganeh, *The Economics of Middle Eastern Oil* (Praeger, New York, 1962)

Jacoby, N., *Multinational Oil, A Study in Industrial Dynamics* (Macmillan, New York, 1974)

Jensen, W., *Energy and the Economy of Nations* (G.T. Foulis, Henley-on-Thames, 1970)

Keenan, P., R. Dobias, M. O'Neil and N. Anderson, *Financial Analysis of a Group of Petroleum Companies* (Chase Manhattan Bank, New York, 1980

Kenen, P.B., *Giant Among Nations, Problems in United States Foreign Economic Policy* (Rand McNally, Chicago, 1960)

Kennedy, W.J. (ed.), *Secret History of the Oil Companies in the Middle East* (2 vols., Documentary Publications, Salisbury, North Carolina, 1979)

Kindleberger, C.P., *American Investment Abroad: Six Lectures on Direct Investment* (Yale University Press, New Haven, and London, 1969)

Kolko, J. and G. Kolko, *The Limits of Power, The World and United States Foreign Policy* (Harper and Row, New York, 1972)

Lawrence, D., *The True Story of Woodrow Wilson* (Doran, New York, 1924)

Lenin, V.I., *Imperialism: The Highest Stage of Capitalism* (International, New York, 1969)

Leontieff, W., 'Theoretical Assumptions and Unobserved Facts', *American Economic Review*, vol. 61, no. 1 (1971)

Levy, W.J., 'An Atlantic-Japanese Energy Policy', *Foreign Policy*, no. 11 (1973)
———, *The Search for Oil in Developing Countries: A Problem of Scarce Resources and Its Implications for State and Private Enterprise* (International Bank for Reconstruction and Development, Washington, 1961)
Lewis, C., *America's Stake in International Investment* (The Brookings Institution, Washington, 1938)
Long-Term Cooperation Programme (International Energy Agency, Paris, 1976)
Magdoff, H., *The Age of Imperialism, The Economics of U.S. Foreign Policy* (Modern Reader Paperbacks, New York and London, 1969)
Manes, R.A., 'Import Quotas, Prices and Profits in the Oil Industry', *The Southern Economic Journal*, vol. 30, (1963)
Marx, K. and F. Engels, *The Communist Manifesto* (International, New York, 1980)
Menderhausen, H., *Dollar Shortage and Oil Surplus in 1949-1950* (Princeton University Press, Princeton, 1950)
Montague, G., *The Rise and Progress of the Standard Oil Company* (Harper and Brothers, New York, 1903)
The National Energy Program (Energy, Mines, and Resources, Canada, Ottawa, 1980)
Nearing, S. and J. Freeman, *Dollar Diplomacy*, reprint (Monthly Review Press, New York and London, 1966)
Noreng, O., 'Friends or Fellow Travellers? The Relationship of Non-OPEC Exporters with OPEC', *The Journal of Energy and Development*, vol. 4, no. 2 (1979)
O'Connor, H., *The Empire of Oil* (Monthly Review Press, New York, 1962)
———, *World Crisis in Oil* (Monthly Review Press, New York, 1962)
O'Connor, R., *The Oil Barons: Men of Greed and Grandeur* (Little, Brown and Company, Boston, 1971)
Odell, P., *Oil and World Power: Background to the Oil Crisis* 3rd edition (Penguin Books, Harmondsworth, 1974)
'The Oil Crisis: Is There a Military Option?' *The Defense Monitor*, vol. 8, no. 11 (1979)
Robertson, N. (ed.), *Origins of the Saudi Arabian Oil Empire, Secret U.S. Documents* (Documentary Publications, Salisbury, North Carolina, 1979)
Robinson, A., *Towards a New Energy Pattern in Europe* (Organisation

for European Economic Cooperation, Paris, 1960)

Robinson, J., 'The Second Crisis of Economic Theory', *American Economic Review*, vol. 62, no. 2 (1972)

Sampson, A., *The Seven Sisters, The Great Oil Companies and the World They Shaped* (The Viking Press, New York, 1976)

Sawhill, J., K. Oshima, and H. Maull, *Energy: Managing the Transition* (The Trilateral Commission, New York, 1978)

Scherer, F.M., *Industrial Market Structure and Economic Performance* (Rand McNally, Chicago, 1970)

Schumpeter, J., *Capitalism, Socialism, and Democracy* (Harper and Brothers, New York and London, 1947)

Shaffer, E.H., *The Oil Import Program of the United States, An Evaluation* (Praeger, New York, 1968)

Solberg, C., *Oil Power* (Mason/Charter, New York, 1976)

Solo, R., 'Gearing Military R & D to Economic Growth' in J.L. Clayton (ed.), *The Economic Impact of the Cold War, Sources and Readings* (Harcourt, Brace and World, New York, 1970)

Stork, J., *Middle East Oil and the Energy Crisis* (Monthly Review Press, New York and London, 1975)

Sucre, L.A., 'The Impact of Multinational Oil Companies on Venezuela: The Oil Companies in General and Gulf Oil In Particular' in J.P. Gunneman (ed.), *The Nation-State and Transnational Corporations in Conflict with Special Reference to Latin America* (Praeger, New York and London, 1975)

Tanzer, M., 'Oil Exploration Strategies: Alternatives for the Third World' in P. Nore and T. Turner (eds.), *Oil and Class Struggle* (Zed Press, London, 1980)

——, *The Political Economy of International Oil and the Under-developed Countries* (Beacon Press, Boston, 1969)

——, *The Race for Resources, Continuing Struggles over Minerals and Fuels* (Monthly Review Press, New York and London, 1980)

Tarbell, I., *The History of the Standard Oil Company* (2 vols, the Macmillan Company, New York, 1933)

Thompson, S., 'Prorationing of Oil in Alberta and Some Economic Implications' (unpublished MA Thesis, University of Alberta, 1968)

Tugendhat, C. and A. Hamilton, *Oil, The Biggest Business* (Eyre Methuen, London, 1975)

Turner, L., *Oil Companies in the International System* (The Royal Institute of International Affairs, London, 1978)

United Nations Economic Commission for Europe, *The Price of Oil in Western Europe* (Geneva, 1955)

US Armed Forces Information School, *The Army Almanac* (US Government Printing Office, Washington, 1950)

US Federal Trade Commission, *The International Petroleum Cartel* (US Government Printing Office, Washington, 1952)

US Fuel Administration, *Final Report of the United States Fuel Administration, 1917-19* (US Government Printing Office, Washington, 1921)

US House of Representatives, Committee on the Judiciary, *Hearings, Nomination of Nelson A. Rockefeller* (US Government Printing Office, Washington, 1974)

US Senate, Subcommittee on Multinational Corporations of the Committee on Foreign Relations, *A Documentary History of the Petroleum Reserves Corporation, 1943-1944* (US Government Printing Office, Washington, 1974)

Vallenilla, L., *Oil: The Making of a New Economic Order, Venezuelan Oil and OPEC* (McGraw-Hill, New York, 1975)

Walden, J., 'The International Petroleum Cartel — Private Power and Public Interest', *Journal of Public Law*, vol. 2, no. 1 (1962)

Wilkins, M., *The Emergence of Multinational Enterprise: American Business Abroad from the Colonial Era to 1914* (Harvard University Press, Cambridge, Mass., 1970)

——, *The Maturing of Multinational Enterprise: American Business Abroad from 1914 to 1970* (Harvard University Press, Cambridge, Mass., 1974)

Williams, W.A., *The Roots of the Modern American Empire: A Study of the Growth and Shaping of Social Consciousness in a Marketplace Society* (Random House, New York, 1969)

——, *The Shaping of American Diplomacy: Readings and Documents in American Foreign Relations* (2 vols, Rand McNally, Chicago, 1956)

Williamson, H.F., R. Andreano, A. Daum, and G. Klose, *The American Petroleum Industry, The Age of Energy, 1899-1959* (Northwestern University Press, Evanston, Ill., 1963)

——, and A.R. Daum, *The American Petroleum Industry, The Age of Illumination* (Northwestern University Press, Evanston, Ill., 1959)

INDEX

Acheson, D. 210, 212
acquisitions: by businesses 5, 12,
   oil companies 27, 33-4, 57, 61-8,
   131, 199, 203-9, *see also* affiliates,
   mergers; US 5-7, of oil reserves
   44, 68, 85, 173, without
   annexation 8, 77-8
affiliates 34-5, 96, 189; *see also*
   acquisitions, mergers
Afghanistan 119-20
Africa 13, 67, 77, 84; North Africa
   163, 164
agriculture 12-15, 36, 55, 70, 83, 146,
   158
Akins, J. 174-86
Alberta 135-9, 216
Aldrich, W. 214, 218
allocations: by Alberta 135, 137;
   by International Energy Agency
   177-9; temporal 231; under US
   oil import programme 128-31;
   see also allowable, conservation,
   prorationing, quotas
allowable 122, 127; *see also*
   allocations, conservation,
   prorationing, quotas
American Empire 2, 37, 82-4, 230
American Independent Oil
   Company 107, 120, 123
Amoco 208; Amoco Mineral 203;
   see also Standard Oil (Indiana)
Anglo-American Oil Co. 33, 52; *see*
   also Standard Oil (New Jersey)
Anglo-Iranian Oil Co. 41, 85, 91,
   104-5, 120; *see also* Anglo-
   Persian, British Petroleum
Anglo-Persian Oil Co. 41, 50, 52, 64,
   67; *see also* Anglo-Iranian, British
   Petroleum
Anglo-Saxon Petroleum Co. *see*
   Royal Dutch Shell
annexation 8, 77; *see also*
   acquisitions
anti-trust 205, 215; action against
   Standard Oil 27; cartel case 106;
   Division, US 106, 113;
   exemptions 107, 178-9; laws 27,
   53, 92, 113, 194; Subcommittee

on 100; *see also* Sherman Act
Arabian-American Oil Co. (ARAMCO)
   66, 90-3, 95-104, 114-19, 189-91,
   210; dispute with Superior 114-19;
   entry of Jersey and Socony 90-3;
   gold royalty 98; oil purchases and
   sales 189-91; Saudi Arabian income
   tax 98-104
ARCO 196-7, 201, 203-4, 206, 209,
   225, 227
Armand, L. 165-6; Armand Report
   151-5
Aruba 60-1, 69, 132
Asia 13, 32-3, 77, 84, 227
Atlantic Refining Co. 46, 53, 59, 72,
   192
atomic energy 144, 151-2; *see also*
   Euratom
automobile 35, 44, 121, 144-5,
   162, 196

Bahrein 63, 67, 69
balance-of-payments: Europe 153-5,
   160; US 10, 80, 169-70
balance-of-trade 65, 79, 81-2, 109
balance-on-goods-and-services 81,
   109
bases 8-9, 42, 77
Bataafsche Petroleum Maatschappij
   54
Blair, J. 100-1, 107, 111-12, 124,
   141, 181, 187
Bolivia 61, 67-8
book value, US investments 12,
   80-1, 150, 173
boycott 62, 176-7; *see also* embargo
Bransky, O. 94-5, 110
Brewster, O. 143
Britain 4-6, 41, 82, 89, 116, 143,
   173, 204, 206; control of oil
   68; conversion to oil 41-3;
   displacement from Middle East
   90-5, 104-5; Iran 105, 116;
   Iraq 50-2; Mexico 46; US
   ambassador to 214-15; *see also*
   England, United Kingdom
British Columbia 203, 206-8
   see also Canada

237